A Colour Atlas of
Virology

A Colour Atlas of
VIROLOGY

Dr. J. Versteeg
Professor of Virology,
Faculty of Medicine,
State University of Leiden,
Head of the Central Clinical Virus Laboratory,
Academic Hospital,
Leiden,
The Netherlands

WOLFE MEDICAL PUBLICATIONS LTD

Copyright © Jan Versteeg, 1985
Published by Wolfe Medical Publications Ltd, 1985
Printed by Royal Smeets Offset b.v., Weert, Netherlands
ISBN 0 7234 0181 0

This book is one of the titles in the series of
Wolfe Medical Atlases, a series which brings
together probably the world's largest systematic
published collection of diagnostic
photographs.
For a full list of Atlases in the series, plus
forthcoming titles and details of our surgical,
dental and veterinary Atlases, please write to
Wolfe Medical Publications Ltd, Wolfe House,
3 Conway Street, London W1P 6HE.

General Editor, Wolfe Medical Atlases:
G. Barry Carruthers, MD(Lond)

All rights reserved. The contents of this book, both
photographic and textual, may not be reproduced
in any form, by print, photoprint,
phototransparency, microfilm, microfiche, or any
other means, nor may it be included in any
computer retrieval system, without written
permission from the publisher.

Contents

Preface	7
Acknowledgements	8
Making a virus diagnosis	9
USE OF PIPETTES	22
FILTRATION	26
WASHING AND STERILISATION	28
Chick embryo techniques	38
CANDLING THE EGG	42
INOCULATION OF ALLANTOIC CAVITY, HARVESTING THE ALLANTOIC FLUID	45
INOCULATION OF AMNIOTIC CAVITY	50
INOCULATION OF CHRIOALLANTOIC MEMBRANE	54
Experimental animals	67
PRODUCTION OF ANTISERA	68
PRACTICAL USES FOR ANTISERA	69
SUCKLING MICE IN DIAGNOSTIC VIROLOGY	70
INOCULATION OF SUCKLING MICE	75
MARKING MICE	79
HANDLING LABORATORY ANIMALS	84
COLLECTING BLOOD FOR ERYTHROCYTES	94
Cell culture techniques	98
TISSUE CULTURE MEDIA	98
MONOLAYERS	99
CELL CONTAMINATION OF CELL CULTURES	99
MICROBIOLOGICAL CONTAMINATION	99
CELL CULTURE SYSTEMS	100
PRIMARY KIDNEY CELL CULTURES	101
MOUSE EMBRYO CELLS	104
CHICKEN EMBRYO FIBROBLASTS	107
PERFUSION TRYPSINIZATION	109
CELL CULTURING METHODS	114
THE COLLODIUM METHOD	120
CELL SLIDES AND PETRI DISHES	123
MICROTITRE PLATE CULTURE	126
SLIDE CULTURE IMMUNOFLUORESCENCE	130
CHLAMYDIAE IN CELL CULTURE	131
Cytopathic effects	133
ASPECTS OF NORMAL CELL CULTURE	134
BACTERIAL AND CELL CONTAMINATION	138
NON-SPECIFIC CYTOPATHIC EFFECT	140
VIRAL CYTOPATHIC EFFECT	142

Histopathology of viral infections	158
DIRECT TECHNIQUES FOR DETECTING VIRUS IN SPECIMENS	171
ELECTRON MICROSCOPY	171
IMMUNOFLUORESCENCE MICROSCOPY	179
HYBRIDIZATION TECHNIQUES	184
RESTRICTION ENDONUCLEASE ANALYSIS OF VIRAL DNA	185
Serological tests	188
PRETREATMENT OF SERA USED IN SEROLOGICAL TESTS	188
FLUORESCENT ANTIBODY TECHNIQUES	209
ELISA TECHNIQUES	212
THE RIA TEST	212
IgM TECHNIQUES	219
MONOCLONAL ANTIBODIES	222
STORAGE OF SERUM SAMPLES AND ERYTHROCYTES	227
LYOPHILIZATION	229
Further reading	233
Index	235

Preface

The expanding demand for virological diagnostic service during the last decennium and the expectation that potent antiviral drugs will be available in the near future has led to the urge to train the necessary manpower.

A number of good textbooks on medical and veterinary virology and some fine manuals on diagnostic procedures for viral infections form a sound basis for the gathering of the necessary knowledge. This Atlas provides a visual introduction to diagnostic virology for those who have to learn the laboratory procedures – microbiologists and pathologists in training, students of medicine, veterinary medicine, medical technology and biological sciences.

The procedures illustrated comprise the use of tissue culture techniques, embryonated eggs and laboratory animals to grow viruses: the identification of viruses, and the antibodies induced by them, with the help of serological techniques. The study of changes in cells and tissues will give insight to pathological processes. Safety rules and laboratory management are interwoven in the chapters. It is clear that these fundamentals are the basis for developing new techniques and for doing research.

As science is rapidly growing and equipment is improving daily, the apparatus illustrated in this Atlas will be changed in due course. What we show are the principal items and not what you ought to buy.

After 30 years in clinical virology I know that this Atlas will also provide desirable information for clinicians on how things are done in the laboratory. Mutual understanding will not only lead to better diagnosis of viral infections, but will result in joint research activities.

Acknowledgements

Because it is impossible for an author to produce all the illustrations for an Atlas like this without help from friends, I am most grateful to the following colleagues for providing me with the following figures:
Dr A.J. van der Eb (**447, 448,, 449, 460, 461, 487, 488**); Dr M. van der Ploeg (**482, 484, 485, 485**); Dr J. Lindeman (**450, 454, 456, 457, 458, 462, 463, 464**); and the late Dr J.C.H. de Man (**455, 459**).

I wish to express my special thanks to Dr F. Eulderink and Dr W.A. van Vloten for the opportunity to use their archives of microscopic preparations, and to Dr L.T.S. van Ekdom and Mr F. Leupe for their technical assistance with the chapter on laboratory animals.

Mr G. Mees was so kind as to make the hundreds of colour prints needed for the layout.

Making a virus diagnosis

The diagnosis of a virus disease can be made in several ways, using direct and indirect methods.

The direct methods, a number of which are called *rapid diagnostic methods*, use the clinical specimen directly to detect the presence of the virus, or to reveal specific changes caused by the action of the virus. The time needed for a diagnosis is two to twenty-four hours. The direct methods are considered to be extremely valuable help for clinical decisions. It must be said, however, that all direct methods must be confirmed by indirect methods as false positives and false negatives are common and because the type of the virus is often not established by rapid methods.

The specimen from the patient has to be taken with care – i.e. at the right time from the right place. Many viruses will only be found in the first few days. Other viruses are extremely susceptible to environmental conditions and will die in half an hour if not transported in a special virus transport-medium or kept at 4°C. Furthermore the kind of sample needed is not always obvious. For example in cases of meningitis it is wise to send a stool specimen as well, as many enteroviruses are involved in meningitis. Clinical specimens are inoculated, after the necessary pretreatment, into cell cultures, experimental animals or embryonated eggs. The culturing methods are time-consuming, at least 24 hours are needed but in most cases it takes a number of days before the manifestation of the virus can be observed, sometimes even up to a few weeks.

Sometimes the diagnosis can be speeded by doing fluorescence microscopy on the cell culture. In animals the viruses can induce clinical signs for which the animals are inspected daily. Cell cultures show cytopathic effects. Embryonated eggs show lesions on the membranes, or the fluid from the amnion sac or allantois sac agglutinates red blood cells.

Serological methods take at least a week to ten days to indicate a significant rise in titre or a diagnostically important rise in IgM. As all patients' samples, even serum, must be considered potentially contagious, good techniques and safe equipment are necessary to prevent accidents. Every worker should be given a list of safety rules and urged to keep to them. Growing viruses means multiplying the original inoculum more than a millionfold and turning it into waste as soon as the diagnosis is made. A safe system for disposing and sterilizing infectious materials is obligatory.

1 Transport of swabs for virological diagnosis. Swabs are to be taken so as to collect the maximum of virus. Always rub the epithelium, as viruses are intracellular. Don't pick up just mucus or secretions. If blisters are present, open the blisters and pick up the material from the bottom. Put the swab in gly-medium to transport it to the laboratory.

2 Stock of gly-medium. Gly-medium is an isotonic buffer, containing minerals, lactalbumin, gelatin, yeast extract and antibiotics. The function of gly-medium in transport of patients' samples is to prevent drying out of the material, to inhibit bacterial growth and to dilute antibodies which may come with the sample.

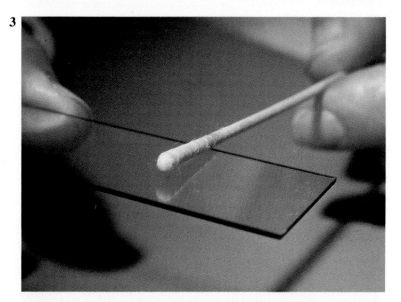

3 Making a slide preparation. Slides for direct diagnosis by staining and immunofluorescence are made by rolling the swab over the glass. In this way cells are stretched on the glass and will be valuable for the diagnosis.

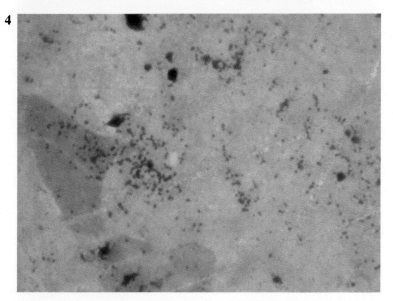

4 Stained slide from an orf lesion. In case the skin lesion is caused by one of the poxviruses, a slide is stained by the Morosow silver stain. In this slide the small dark dots are virus particles from the skin lesion of a farmer infected with orf (sheep pox).

5 Making an impression smear. Biopsy material can be used by making cryostate section or impression smears. In this case a piece of brain tissue was held on a wooden spatula and a microscope slide was pressed against the tissue to pick up cells.

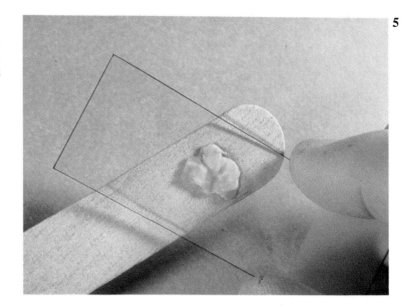

6 Immunofluorescence on an impression smear. The impression smear was made as shown, fixed in aceton and stained for the presence of herpesvirus. Both cells show intranuclear and cytoplasmatic fluorescence typical for herpesvirus.

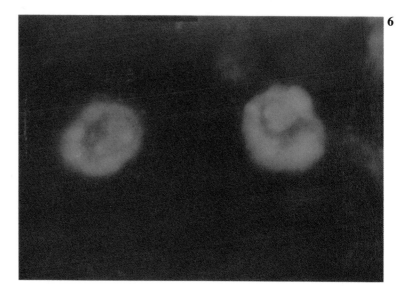

7 Blood smear mononucleosis. A blood smear showing atypical lymphocytes is indicative of mononucleosis although differentiation from the closely-resembling cytomegalovirus-induced disease must be done serologically.

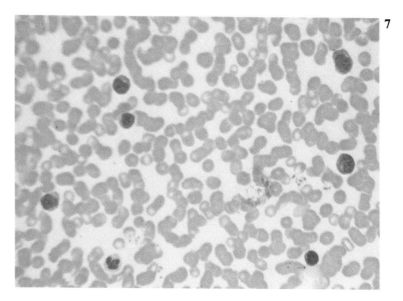

8 Sending slides by post. The dispatching of slides, necessary for consultation, special staining or immunofluorescence, brings the risk of breakage. A number of rigid plastic containers can be obtained from firms specializing in laboratory supplies. None of these holders offers complete protection; shipment will be safer when additional protection is provided by using a cardboard box.

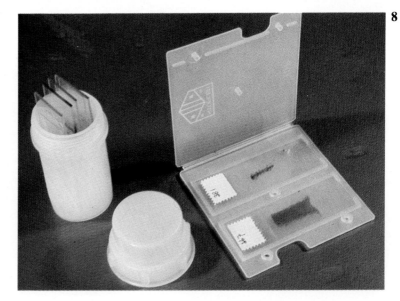

9 Stool sample. Screw cap bottles are preferred for shipping faecal samples. A small piece, as large as a big hazelnut, is sufficient. Do not stuff the bottle as this easily contaminates the outside and if the faeces ferments during shipment the contents of the bottle will be expelled by pressure on opening.

10 Making a faecal suspension (1). A 20% suspension of the faeces is made by weighing about 1 gram of faeces and adding 4 millilitres of saline. Shake in a jar with glass beads. The suspension is then centrifuged at 3500 r.p.m. to clear.

11 Making a faecal suspension (2). The bottle with the glass beads and the faeces are shaken mechanically. As an aerosol is formed over the suspension it is safer to let the bottle stand for 5 minutes, to settle down the aerosol, before opening.

12 Urine sample. Urine samples are collected as a 'mid-stream' portion, as absolute sterility is not necessary. After washing the genitals the first portion of urine is voided and the beaker is held in the stream to collect a sample. Sometimes the urine sample is buffered by adding a solution like tris buffered hanks. Viruses recovered from urine are: mumps-, measles-, papova-, cytomegalovirus.

13 Blood samples. Blood samples must be collected in such a way that no blood is spilled on the outside of the tube. Blood must always be considered as a contagious material.

A fine system of collecting blood is the vacuum system, in which the blood sample is drawn into an evacuated tube.

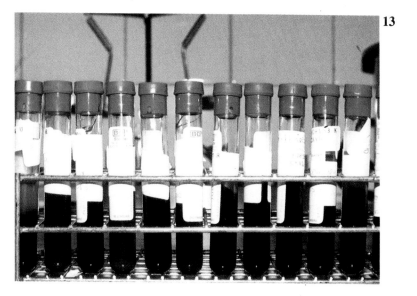

14 The quality of the serum sample. The quality of the serum can be judged from its appearance. Left to right:

1 Normal serum, taken before a meal, is translucent
2 Slightly opalescent serum, containing fat. The blood sample was taken some time after a light meal
3 Very turbid serum from blood taken after a copious meal
4 Normal serum
5 Slightly haemolytic serum, the blood sample was kept over the weekend
6 Strongly haemolytic serum from a blood sample that had been in the mail for 5 days

15 Sayk sedimentation chamber. When cells from fluids are to be examined for the presence of virus or virus inclusions slides have to be made. An elegant way to do this is to use a Sayk sedimentation chamber in which the cells form a sediment by natural gravity, aided by filter-paper. Always disinfect the Sayk chamber after use.

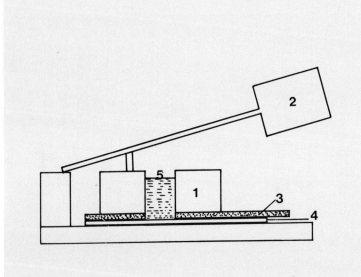

16 Diagram of a Sayk chamber.

1 A plastic ring
2 Weight pressing down the plastic ring
3 Filter paper perforated
4 Microscope slide
5 The fluid to be examined

The fluid is sucked up by the filter paper and cells form a sediment on the glass. The time taken to make a slide is about half an hour.

17 Sedimentation of cells on a microscope slide. The cell sediment on the microscope slide is fixed before staining with dyes or fluorescent antibody techniques. Preparations made by the Sayk method always show undistorted cells.

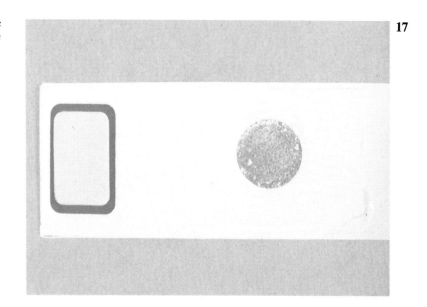

18 The cytocentrifuge. The cytocentrifuge enables us to make cell sediments from fluids on a microscope slide. The cells are deposited on the glass by centrifugal force and the fluid is absorbed into filter paper. Do not use the cytocentrifuge for contagious materials as aerosols are produced. Clean and disinfect the cytocentrifuge after use.

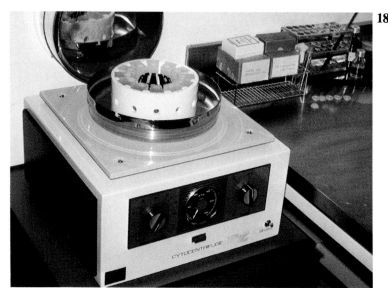

19 The cytocentrifuge slide. The area of the cells deposited is smaller than in the Sayk chamber. However the speed and ease with which the slide is made may be important.

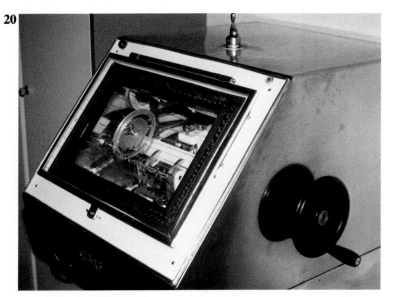

20 Refrigerated microtome. To make cryostate sections the snapfrozen piece of tissue is placed on the chuck of a refrigerated microtome and cut into sections of the desired thickness. The microtome pictured here is enclosed by a housing that is cooled to the level of −30°C to −70°C. All knobs for controlling the microtome are on the outside. Care must be taken to disinfect the knife and the chuck after use; viruses can survive a long time in low temperatures.

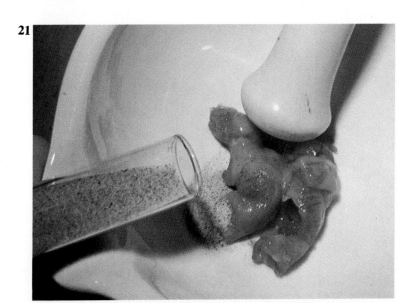

21 Grinding tissues in a mortar. Tissues from man and animal must be ground and extracted before inoculating into tissue cultures, eggs or animals. Grinding is aided by adding acid-washed sterile sand. The work should be done in a safety cabinet as aerosols are produced by this method.

22 Grinding tissues in a homogenizer (1). The beakers of the blending apparatus contain revolving cutting tools. The lid closes airtight to prevent aerosols from spreading. It is good practice to wait 10 minutes after the homogenizing process is ended to allow aerosols to settle down before opening the beaker.

23 Grinding tissue in a homogenizer (2). The speed of the homogenizer can be as high as 16000 r.p.m. This will lead to considerable heating of the material. At higher speeds the beaker has to be placed in ice to prevent loss of viability of the virus.

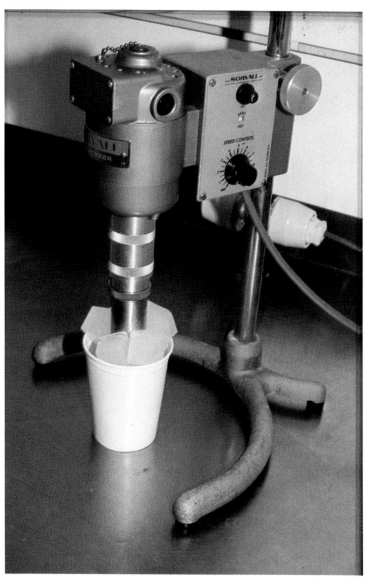

24 Grinding cells. Very fine grinding of small amounts of cells is best done in a special tissue-grinder made of a glass tube in which a teflon plunger fits closely. Slowly turning the shaft and moving it up and down will break up all the cells.

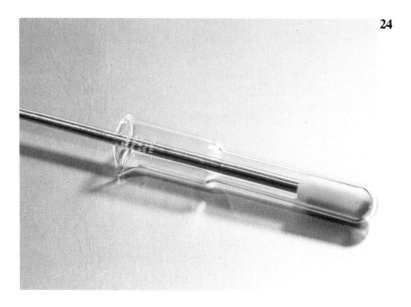

25 Ultrasonic disruption of cells. Sonification of cells can be used for preparing virus antigens, especially for freeing viruses from cell material. Ultrasonic disruption causes a rise in temperature of the material, so adequate cooling is needed. Prevent damage to your ears by wearing protectors.

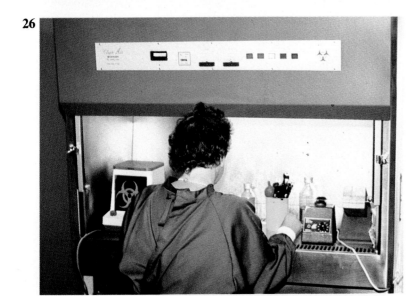

26 Safety cabinets. All work with contagious material has to be done in a safety cabinet to prevent laboratory-acquired infections. Safety cabinets for virological purposes are constructed so that all aerosols are carried away over a filter before the air is disposed of. The worker is protected by a curtain of sterile air.

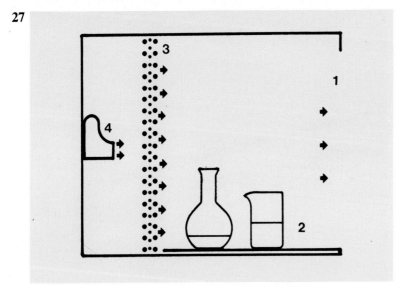

27 Laminar flow clean air cabinet. This protects the work against microbial contamination by a laminar flow of sterile air. This type of cabinet is suitable for making media and handling non-infected tissue cultures. It is not suitable for infectious work as personnel are exposed to the exhaust air.

1 Open front and exhaust air
2 Working space
3 Hepafilters and laminar flow
4 Fan

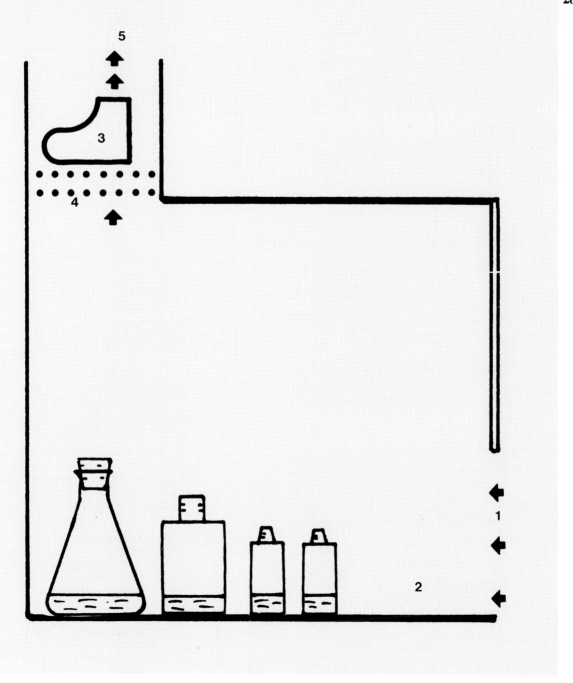

28 Class I microbiological safety cabinet. This protects the worker. Air is drawn from the room, with the air-stream away from the worker thus protecting him against infection. The incoming air is not sterile and may contaminate the work. The exhaust air is filtered to protect the environment.

1 Open front with inward airflow
2 Working space
3 Exhaust fan
4 Exhaust hepafilter
5 Sterile exhaust air

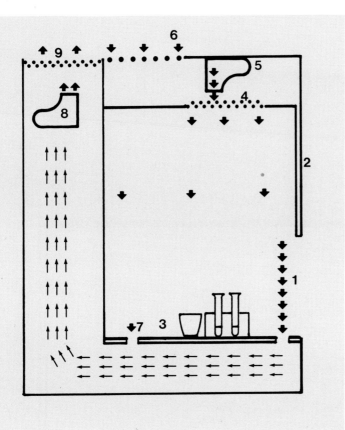

29 Class II microbiological safety cabinet.
This protects both the work and the worker. Incoming and exhaust air is filtered. The open front is protected by an air current, which acts as a curtain. Highly contagious material must be handled in a Class III cabinet as air and liquid spilling through the front are not absolutely preventable.

1 Open front with air curtain
2 Window
3 Working space
4 Inlet with hepafilter
5 Inlet fan
6 Air inlet with prefilter
7 Air vents in the bottom
8 Exhaust fan
9 Exhaust hepafilter

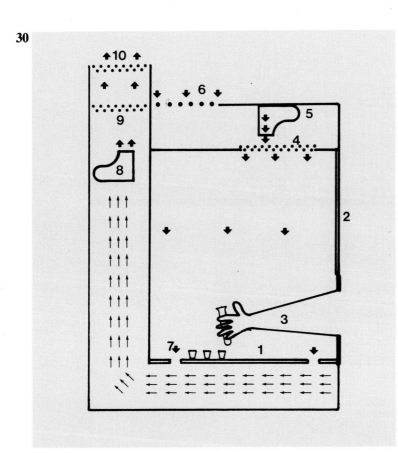

30 Class III microbiological safety cabinet.
Maximal protection for the test, the worker and the environment. Regular inspection of the filters and the fan motors is necessary for maintaining the maximal safety level.

1 Working space
2 Window
3 Gloves, making the cabinet front totally closed
4 Air inlet with hepafilter
5 Air inlet fan
6 Air inlet prefilter
7 Air vents to duct in the bottom
8 Exhaust fan
9 and 10 A double set of hepafilters in the exhaust

31 Tissue culture inoculation. Materials from patients are inoculated into tissue cultures to assay them for the presence of viruses. This work is done in a safety cabinet to protect the culture from being contaminated and to prevent infection of the worker.

A tissue culture in which there is optimal virus replication contains up to 10^8 viable virus particles.

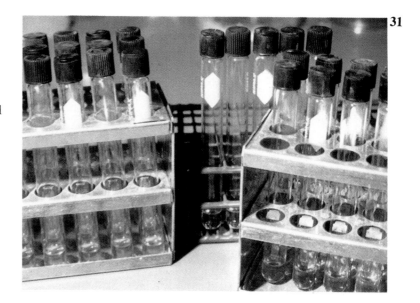

32 Cytopathic effect in tissue culture. Virus replication will cause pathological changes and subsequent cell death. These changes – observed in the living culture – are called cytopathic effects (cpe). In a stained slide the nature of the cytopathic effect can be observed, such as cell retraction, rounding of cells, giant cells, inclusion bodies, and so on.

Typing of the virus is done by serum neutralization.

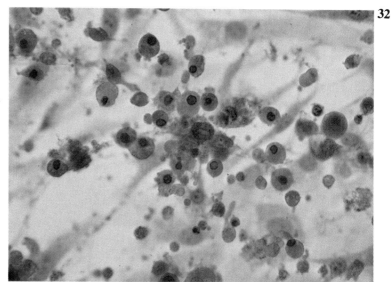

33 Fluorescent microscopy for rapid diagnosis. When rapid diagnosis is required the inoculated tissue culture can be submitted to immunofluorescent staining even when the cpe is not yet visible. Using a type-specific serum the diagnosis can be made within one or two days after inoculation.

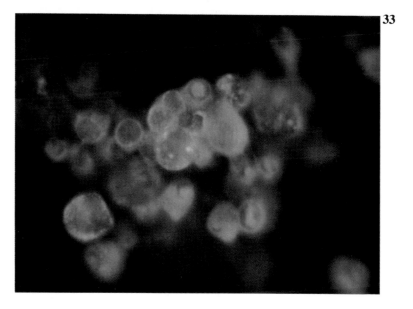

USE OF PIPETTES

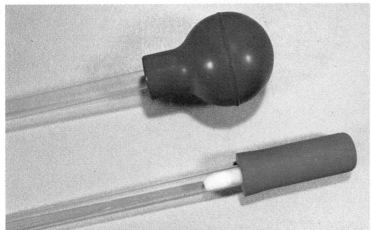

34 Pipette balloons (1). Mouth pipetting is dangerous and should be prohibited. Simple balloons can be used with non-graduated pipettes for simple work like removing the supernatant after centrifugation.

Do not use the bulb to blow out the pipette vigorously; this causes aerosols.

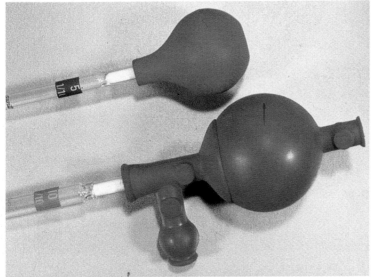

35 Pippette balloons (2). Simple balloons can be used with graduated pipettes but are hard to use in this way. Special balloons with valves offer all the functions needed. The three valves function as follows: the top valve is used to evacuate the balloon, the lower valve to fill the pipette by suction through the evacuated balloon, the side valve lets in air to empty the pipette.

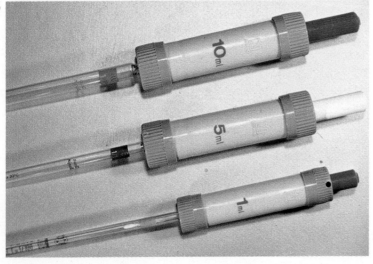

36 Pipette pumps (1). Simple pipette pumps are used with calibrated pipettes to draw the fluid in. The amount in the pipette is regulated by letting air in at the top of the pump.

37 Pipette pumps (2). More sophisticated pumps are more comfortable to use. The drawing and expelling of the fluid is done by turning the wheel. When distributing equal amounts to tubes the valve can be used to let in air to drive the fluid out.

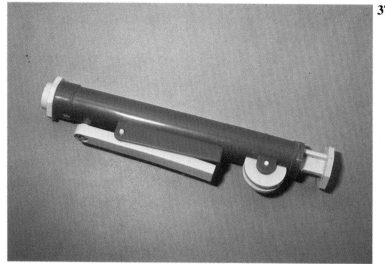

38 Eppendorf pipettes. The Eppendorf system uses airpumps with disposable tips. They are especially suited for working with small amounts of fluid. The tips are disposable and can be gas-sterilized for tissue culture work.

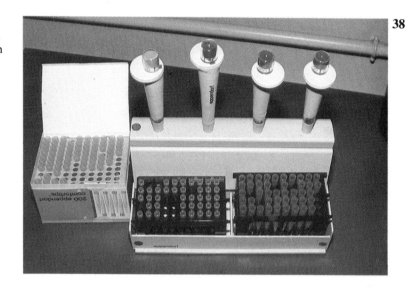

39 Sterilizing glass pipettes. Graduated and non-graduated pipettes are plugged with cotton wool and sterilized by dry heating at 160°C. Stainless steel boxes or glass tubes can be used as containers. Individual pipettes can be paper-wrapped and dry sterilized or plastic-wrapped and gas-sterilized.

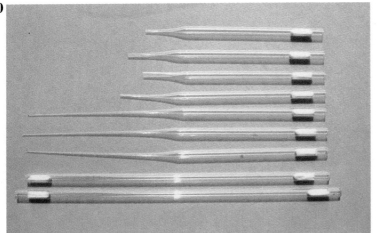

40 Pasteur pipettes. Pasteur pipettes are made by drawing out a glass tube. When closed with a cotton plug, the tubes can be sterilized before drawing out.

The pasteur pipette is used with a rubber bulb to handle small amounts of non-measured fluids, for example for removing serum from a clotted blood sample or changing medium in tissue culture tubes.

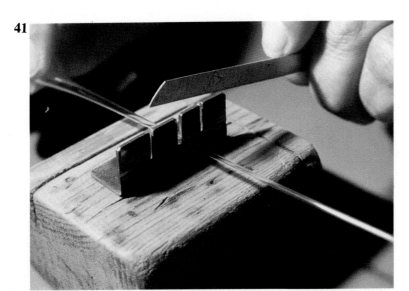

41 Standardizing pasteur pipettes. Pasteur pipettes can be used for quantitive work when they are cut to give a standard drop. This is done by a device called a 'guillotine', made of a piece of copper plate in which slits of determined width are cut. The pasteur pipette is pushed in a slit and cut off at this exact diameter.

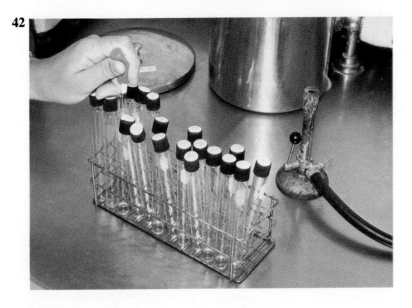

42 Using a pasteur pipette. Pasteur pipettes are useful for changing the medium in individual tissue culture tubes. A simple balloon is used to draw all the fluid in at once. Pipette and fluid are disposed of in the steel bucket filled with disinfecting fluid, seen in the background.

43 Wrong way to use a pipette. Never expel the pipette into the liquid with force. This causes aerosols which can be infectious or can contaminate neighbouring tubes.

44 The right way to use a pipette. Hold the tip of the pipette against the wall of the vessel and empty the pipette slowly. In this way there will be no drops hanging on the pipette tip when the fluid is released and no aerosol is formed.

45 Releasing the balloon from the pipette. Releasing the balloon from the pipette causes aerosols. To prevent this immerse the tip of the pipette under the disinfectant before taking off the balloon.

FILTRATION

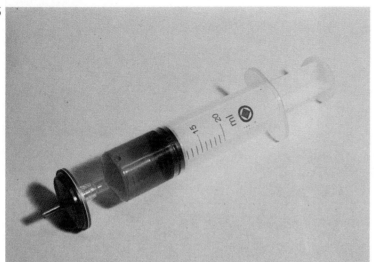

46 Filtering small amounts. Small amounts of fluids can be made bacteriologically sterile by filtering through small disposable filtering units, which can be connected to a disposable syringe.

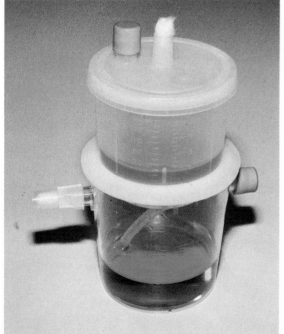

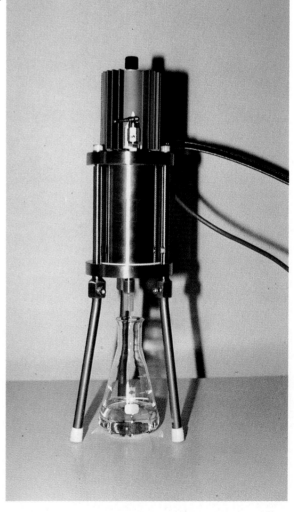

47 Filtration of larger amounts. Larger amounts of fluids like medium, serum and buffers can be sterilized by filtration. Up to 100 ml is filtrated in disposable units, larger amounts in reusable units which must be sterilized before use. All these filters will let viruses pass.

48 Filtration to separate viruses and proteins. When viruses and proteins must be removed or concentrated from fluids, special fine pore filters are used. These require high pressures to pass the filter, and agitation to prevent sludging of the filter. These filters are particularly useful for making antigens.

49 Centrifuges. For safety reasons centrifuges must be firmly closed with no moving parts on the outside. A microswitch will cut off the current to stop the rotor when the lid is opened before the end of the centrifugation process.

50 Balancing the buckets. Balancing the buckets to prevent breaking of the centrifuge tubes and the production of aerosols is important. If a tube is broken, keep the lid on the centrifuge closed for 10 minutes, to let the aerosol drops settle, before inspecting the contents.

51 Closing centrifuge tubes. Open centrifuge tubes always cause aerosols. Glass tubes can be closed by folding the cotton plug and securing with a rubber band or a rubber stopper or with paraplast film.

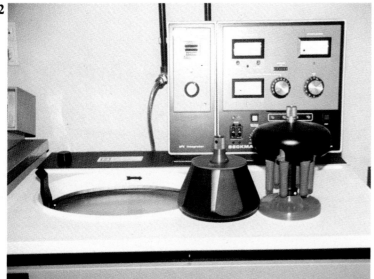

52 Ultracentrifuge. The ultracentrifuge is used at high speeds – 45000 to 60000 r.p.m. – to sediment viruses for purification or concentration. Antibodies can be separated by centrifugation in sucrose gradients. Different types of rotors are used for the various functions.

WASHING AND STERILIZATION

53 Washing and disinfecting hands. It is important to wash the hands after each job. Use a faucet that can be turned on and off with a pedal or an elbow handle. For disinfection povidon-iodine soap is used, which is dispensed from a bottle that is controlled by a pedal or elbow lever.

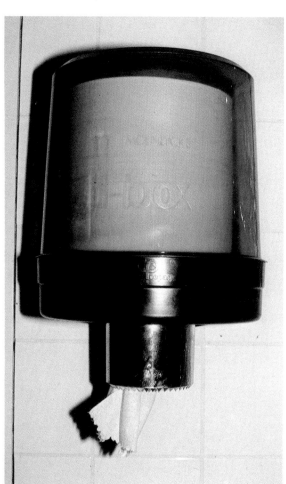

54 Paper towels (1). Paper towels often come in dispensers with single sheets which are relatively hard to get out. The metallic parts will then be smeared and contaminate the hand of the next user.

55 Paper towels (2). The system in which the paper of the towel is on a roll and the pieces needed for individual use are easily pulled out and torn off. This system is the best for containment.

56 Collecting contaminated materials (1). All contaminated materials are collected in stainless steel buckets closed with a lid. At the end of the workday the buckets are collected and sterilized by autoclaving. The contents are washed after cooling down.

57 Collecting contaminated materials (2). Pasteur pipettes, plastic pipette tips, syringes, needles, ampoules, cotton swabs and so on are collected in a stainless steel pan partially filled with a layer of a disinfectant and closed with a lid. After sterilization the contents are discarded.

58 The autoclave. All biologically contaminated material, waste, disposables and reusable glassware and instruments are decontaminated by steam sterilization in the autoclave. All living micro-organisms, including bacterial and fungal spores, are killed at 121°C within 20 minutes.

59 High pressure steam sterilization. Small amounts can be sterilized in the laboratory by using a pressure cooker which is easy to operate. Only those things which can be penetrated by steam can be sterilized this way. Dry material in tightly-closed containers will not be rendered sterile.

60 Control of autoclave sterilization. As incomplete elimination of air and careless packing cause insufficient sterilization, rigid controls must be made. Almost all systems are based on discoloration by heat. The autoclave tape shown here is used on the plastic lids of bottles. To allow penetration of steam the lids are tightened after sterilization.

61 Washing used labware. After sterilization of the stainless steel buckets, containing the used labware, the contents are sorted, freed from stickers and other loose material and washed in a special washing machine. The detergents used are more aggressive than those used in households. The glassware is rinsed in demineralized water to get rid of residues of the detergent.

62 Washing glassware (incorrect method). Glassware used for saline, buffers and other non-adhering fluids is often washed in the laboratory. Wash it with tapwater and rinse it with distilled or demineralized water to prevent drying spots. Do not dry the glass the way pictured because unnecessary crowding of the bench hinders work and causes breakage.

63 Washing glassware (correct method). Use a drying rack for laboratory-washed glassware. Clean the rack and the pegs regularly to prevent contamination of the glass.

64 Ultrasonic cleaning. To clean small complicated objects an ultrasonic cleaning bath can be employed. The sound waves loosen all adhered substances. However only objects uncontaminated by viruses are to be handled as the process creates an aerosol that can spread infection.

65 Dry sterilization. Dry sterilization is done in a specially constructed dry oven, which is heated electrically to a temperature of 160°C for 1 to 2 hours. The hot air is pumped round by a turbine ventilator to ensure even dissipation of the heat. Never pack the oven tightly as this can lead to insufficient sterility of a part of the contents.

66 Dry sterilization. It is important to follow the sterilization instructions carefully and to use controls. The sterilization time starts at the moment that the desired temperature is reached in the oven. Keep the door tightly shut. Be sure that the air turbine is functioning. If a power failure interrupts the sterilizing process let the oven cool down and start the sterilization procedure again for the full time, regardless of the duration of the first period.

67 Control of dry sterilization. In order to verify if the temperature in the oven reached the needed level, tubes with chemical indicators can be suspended in the inside. The chemicals melt and discolour at the sterilization temperature.

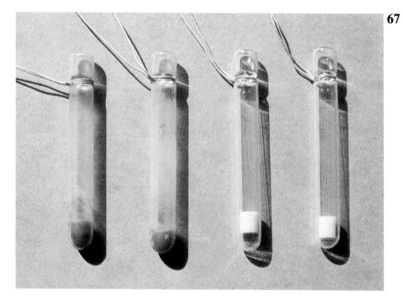

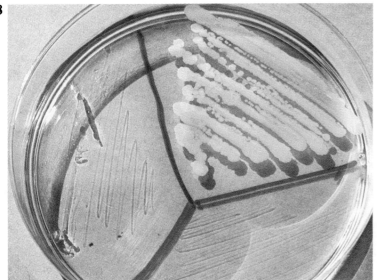

68 Control of dry sterilization. Tubes containing bacterial spores can be placed between the wares to be sterilized. After the sterilization is completed the contents of the spore tube are inoculated on an agar plate. The plate shown here was inoculated with a control tube (bacterial growth) and with two adequately sterilized tubes (no growth).

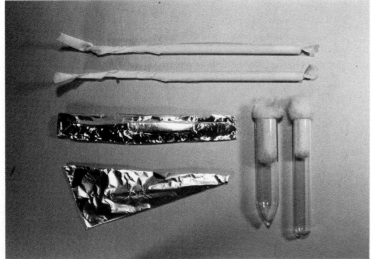

69 Packing for dry sterilization (1). Glassware, porcelain and metal instruments must be wrapped to insure sterility after hot air sterilization. Paper and aluminum foil withstand the heat and are the most used materials for this purpose. Glass tubes can be plugged with unbleached cotton wool.

70 Packing for dry sterilization (2). Mortars and pestles are wrapped in brown paper to sterilize in a hot air stove. As the packing is not air tight absolute sterility will not be maintained forever. Store for a limited time, and stamp a date.

71 Packing for dry sterilization (3). Flasks, bottles and measures must be sterile on the inside, thus it is adequate to cover the opening to guarantee this sterility. Different methods can be followed, each having its own merits.

72 Aluminium foil caps. The openings of glassware can be closed for sterilizing by a cap of folded aluminium foil. A good household quality will do. As the foil will not prevent air coming in and out, due to variations in atmospheric pressure, sterility will not last longer than a few days.

73 Cotton wool plugs. Plugs of cotton wool are excellent for closing glassware for sterilization. However some precautions must be taken. The glassware must be dry prior to plugging and sterilizing in hot air, otherwise the oils of the cotton will come out and will form a film on the glass. Using a piece of gauze around the plug reduces the contamination of the glass with fibres.

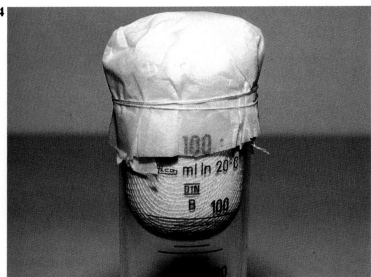

74 Cotton wool plugs and paper. As the top of cotton wool plugs can gather dust it is wise to cover them with paper before sterilization for longer storage.

75 Packing for dry sterilization (4). It is impractical to plug tissue culture tubes for hot air sterilization as these tubes are used in quantities. Besides, substances evaporating from the cotton wool would render the glass surface less fit for cell growth. The tubes are best packed in aluminium boxes that are wrapped in packing paper. Re-sterilise the box with tubes when only a part of the contents was used in the laboratory.

76 Centrifuge tubes. Centrifuge tubes are plugged with unbleached cotton wool covered by cotton gauze. Protect the thick walled glassware from mechanical stress by packing them loosely in a metal basket during hot air sterilization.

77 Gammaray sterilization. Most articles which are mass produced and sold pre-sterilized are sterilized by gamma irradiation. This method will not leave any residues and is therefore ideal for tissue culture wares.

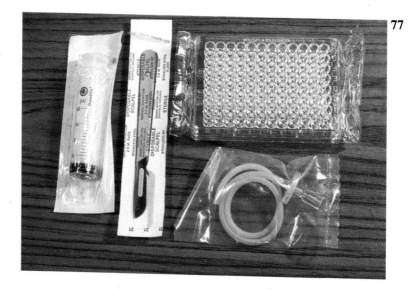

78 Gas sterilization. Plastics and pressed glassware which will not withstand heating can be sterilized by exposure to ethylene oxide gas. As this gas is very toxic the method is not suitable for use in small laboratories which do not have the proper equipment. Most large hospitals have a gas sterilization unit which will provide the service to all who need it.

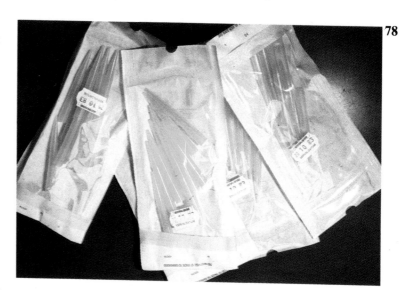

Chick embryo techniques

A large number of human and animal viruses can be propagated directly, or after adaptation, in the different structures of the embryonated egg. The equipment needed for these techniques can be relatively simple and the skill can be learned in a reasonable time.

The advantages of the use of the embryonated egg in the virus laboratory are many. Fertile eggs can be had anywhere in the world and the incubation of the eggs is a well known technique. The different structures of the embryonated egg can support a number of viruses. The inoculation of eggs can be performed aseptically, even in primitive laboratories. After incubation of 8–11 days the embryonated egg can be inoculated. The virus replication takes place in 2–3 days, exceptionally in 5–7 or more days. The contents of the eggs are harvested and inspected for the presence of virus. A number of viruses will induce morphological alterations at the site of inoculation, other viruses will kill the embryo or give rise to the production of haemagglutinins. The production of virus in the infected embryonated eggs is high so that this method is suitable for the production of virus for the preparation of vaccines and of antigens for diagnostic purposes. White leghorn eggs are used by most laboratories because they are easily available but other races of chickens can be as suitable except that the coloured shell is a disadvantage in transillumination. In a number of cases duck eggs are used. An important point is the quality of the egg used. The best source of eggs is a well-run poultry farm, keeping flocks free from common infections and able to guarantee an all year round supply of eggs with a fertility rate of 80–90%. The eggs should also be delivered in a clean condition. Even when all these conditions are fulfilled, the embryonated eggs can still occasionally harbour infectious agents like viruses (Rous sarcoma virus, avian leukosis virus, adenovirus) and bacteria or bacteria-like organisms (salmonella, mycoplasma, chlamydiae).

Check at the poultry farm whether antibiotics are added to the diet of the chickens when the eggs are to be used for work with chlamydiae, mycoplasma and rickettsiae, as these organisms can be inhibited by antibiotics. Although the chick embryo does not produce antibodies the eggs contain a fraction of maternal antibodies. This can be a disadvantage when the incubated eggs are to be used for cultivation of avian viruses.

Pre-incubated eggs can be ordered from commercial hatcheries, so that they can be inoculated immediately at arrival. This can be an advantage when the workload is irregular.

When the eggs are ordered fresh they should preferably be incubated within 5 to 6 days. Eggs that are not used immediately should be stored in a cool place (15–18°C). It is most convenient to incubate the eggs in a commercial, mechanical, egg incubator as these machines operate automatically, maintaining the required temperature and humidity and turn the eggs a sufficient number of times per day. When a special incubator is not available the eggs can be incubated in a conventional bacteriological incubator with provisions made for adequate ventilation and humidity (40–70%). Mostly the temperature range 37.5–38.0°C is chosen for the first incubation period. The eggs must be turned by hand at least twice a day, preferably 4 times.

After 3–5 days of incubation the eggs are transilluminated for the first time with an egg-candling lamp. All non-fertile eggs and those containing a dead embryo are discarded. The second candling is done prior to inoculation at 8, 10, 11 or 12 days depending on the type of virus and the route of inoculation. During the second candling the air space is marked off with a pencil and a suitable site is picked where the egg shell can be drilled in order to inoculate the structures of the chick embryo. For the inoculation of the amniotic cavity, the allantoic cavity and the yolk sac, a small hole is drilled.

For the inoculation of the chorio-allantoic membrane, an equilateral triangle is marked in pencil at the site where no important blood vessels are running. The eggshell is cut through with a rotating carborundum disc taking great care not to perforate the shell membrane. When larger openings are made, mostly at the blunt side, more direct manipulation of the embryo and its structures is possible.

After inoculation the opening is sealed and the eggs are incubated stationary at the same or a lower temperature during a 2–9 day period depending on the type of virus and the inoculation route. To harvest the virus-infected fluids and structures of the egg the shell has to be opened with great care to prevent unwanted dissemination of the virus. Always use safety cabinets for virulent and dangerous viruses. Never use a drill or a carborundum disc to open infected eggs as aerosols are created by these devices. The eggs are opened with a pair of forceps or with scissors. Consider all shell fragments to be contaminated.

When finished, disinfect the bench and autoclave all collected residues in a stainless steel bucket. Prior to harvesting the allantoic or amniotic fluids, the eggs are chilled at 4°C for 4–6 hours, not to kill the embryo per se, but to stop the blood circulation as contamination of the fluids with erythrocytes lowers the virus content of the harvest.

If the embryo, yolk sac or chorioallantoic membrane are to be harvested chilling is undesirable.

Harvested fluids are tested in the haemagglutination test. Positive fluids can be identified by haemagglutination inhibition tests. Inoculated chorioallantoic membranes are harvested and, after washing in saline, inspected for lesions. The lesions can be typical of the virus as is the case in herpes virus hominis type I and II viruses, vaccinia virus, cowpox virus and some arboviruses. In other cases the lesions are just indications of virus multiplication. Non-specific lesions can occur when irritating suspensions are inoculated on the chorioallantoic membrane. Rough handling can cause traumatic ulcera. The harvested membranes can be ground, suspended in buffer and used as virus stock for further work. Death of the embryo can be due to different causes. Early death is generally due to inoculation trauma to which young embryos are especially prone. Another cause of early death is secondary infection with bacteria, sometimes, but not always, accompanied by a foul smell. Bacteriological controls are made in case of doubt.

Death due to the inoculated virus is important as this can be pathognomonic. For example, titration of Japanese B encephalitis can be performed in 10-day-old eggs by the yolk sac route, being a more sensitive method than the usual intracerebral inoculation in mice. The virus kills the embryos in 48–96 hours, as can be seen by daily candling. In some cases the embryo is harvested and parts of it used as a virus source. For example the brain in egg adapted rabies virus or the lungs in mycoplasma pneumoniae.

Smears from the intected sites of the chorioallantoic membrane and the yolk sac can be stained directly. This can be of diagnostic value, for example in the case of the Morosow stain for pox viruses or the Macchiavello and Gimenez stains for elementary bodies of chlamydiae.

Virus	Isolation/Adaptation	Pre-incubation	Inoculation route	Second incubation	Optimal temp	Lesions C.A.M.	Haemagglutination	Death of embryo
Chlamydiae	I	7 days	yolk sac	13 days	35°C	–	–	–
Influenza A and B	I	10–11 days	amniotic	2–3 days	33–34°C	–	+	–
Influenza C	I	7–8 days	amniotic	5 days	33–34°C	–	+	–
Influenza A and B	A	10–11 days	allantoic	2–3 days	34–37°C	–	+	–
Mumps	I	7–8 days	amniotic	5–7 days	35–37°C	–	+	–
Mumps	A	7–8 days	allantoic	5–7 days	35–37°C	–	+	–
Newcastle disease	I	9–11 days	allantoic	2–4 days	35–37°C	–	+	(+)
Newcastle disease	A	9–11 days	c.a.m.	3–6 days	35–37°C	+	+	+
Sendai (paraflu 1)	A	10–11 days	allantoic	2–3 days	35–37°C	–	+	+
Vaccinia	I	10–11 days	c.a.m.	2–3 days	37–38°C	+	–	+
Variola	I	10–11 days	c.a.m.	3–4 days	35–36°C	+	–	(+)
Cowpox	I	10–11 days	c.a.m.	2–3 days	37–38°C	+	–	+
Ectromelia	I	10–11 days	c.a.m.	3–4 days	35–36°C	+	–	(+)
Herpes hominis type I	I	10–11 days	c.a.m.	3 days	35–37°C	+	–	(+)
Herpes hominis type II	I	10–11 days	c.a.m.	3 & 7 days	35–37°C	+	–	(+)
Aujeszki	A	10 days	c.a.m.	3–7 days	35–37°C	+	–	–
Herpes B	I	10–11 days	c.a.m.	3 days	35–37°C	+	–	–
Inf. laryngotracheitis	I	10–11 days	c.a.m.	3–5 days	35–37°C	+	–	–
C.E.E.	A	10–11 days	c.a.m.	3 days	35–37°C	+	+	–
E.E.E./W.E.E./V.E.E.	I	10–11 days	any route	1 day	35–37°C	–	–	++
Yellow fever	A	12 days	c.a.m./embryo	3–4 days	35–37°C	–	–	(+)
Japanese encephalitis	I	11 days	amniotic	2–3 days	35–37°C	+	–	++
St Louis encephalitis	I	10–12 days	yolk/c.a.m.	3–4 days	35–37°C	+	–	+
West Nile	I	10–12 days	yolk/c.a.m.	3–4 days	35–37°C	+	–	++
L.C.M.	A	10–12 days	c.a.m.	3 days	35–37°C	+	–	+
Rabies	A	7 days	yolk	9–10 days	36.5°C	–	–	–
V.S.V.	I	7 days	c.a.m.-all.	1–3 days	35–37°C	+	+	+/++

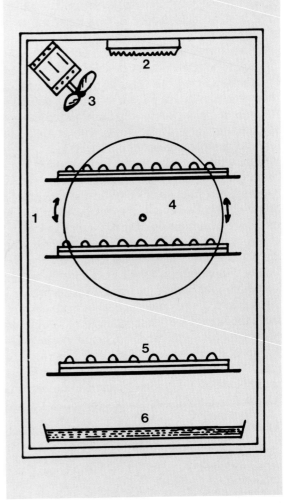

79 The egg incubator. Various incubators are available, and all work on similar principles. In this machine, the eggs are pre-incubated on racks which are rotated automatically every hour (the device at the top and right side). The temperature and humidity can be checked through the window in the door.

80 The egg incubator, schematic diagram.

1 Insulated wall
2 Heating element and thermostat
3 Ventilator
4 Rotating egg racks for pre-incubation
5 Stationary racks for inoculated eggs
6 Water evaporator

81 The egg incubator, inside view. The pre-incubation racks are situated in the upper part of incubator. To encourage normal development the racks are tilted, alternately 45 degrees forwards and 45 degrees backwards every hour. At the bottom the inoculated eggs are incubated in stationary racks.

82 Dating the eggs. The eggs are stamped with the date on which incubation began. The blunt end of the egg is always used to avoid interference with candling.

83 The candling lamp. The candling lamp contains a strong incandescent lamp or a high pressure mercury lamp encased in a light-proof box. In the front of the box is an oval hole where the egg fits for examination. A viewing hood, black on the inside, fits over the viewing hole to enhance vision.

84 The candling lamp, schematic diagram.
1 Wall of the light-proof box
2 High pressure mercury lamp
3 Ballast coil for the lamp
4 Viewing hole for the egg
5 Viewing hood

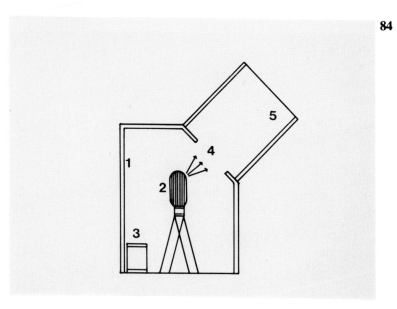

CANDLING THE EGG

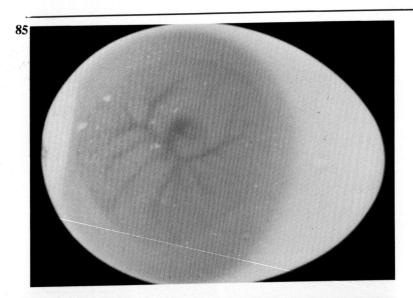

85 Candling the egg (1). The first candling is carried out on the third day of incubation and shows whether the egg is fertile. If not, it is discarded. The air sac is small.

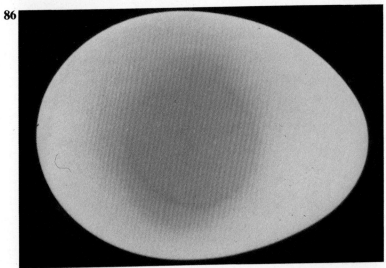

86 Candling the egg (2). Early death of an embryo. When good quality eggs are used, 3 to 8 per cent will not hatch and 2 to 4 per cent will die early. In the latter case a careful inspection will reveal a small shadow indicating the dead embryo.

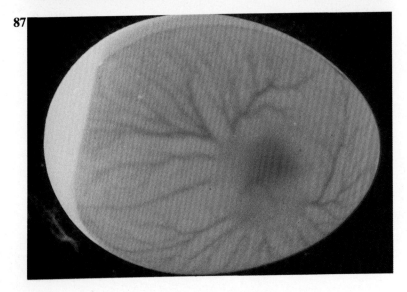

87 Candling the egg (3). At 5 days the embryo can be seen clearly. The air sac is visible and will become larger during incubation due to loss of water vapour.

88 Candling the egg (4). This 6 day-old egg shows a number of small cracks in the shell due to careless handling. These cracks accelerate the evaporation of water, as the very large air sac indicated. In most cases these eggs are discarded.

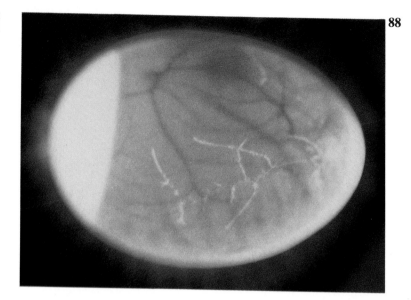

89 Candling the egg (5). Day 8 is the ideal time for inoculation of the amniotic sac if mumps virus is suspected. The amniotic cavity is clearly shown and the pigmented eye of the chick embryo indicates the position of the head.

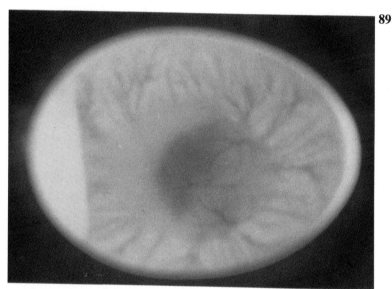

90 The contents of a 10-day-old incubated egg. The structures inside the egg are quite dense and orientation is difficult when candling. The majority of inoculations are made in 10 to 11-day-old eggs. The embryo is well developed and lays in the amniotic sac which contains 2 ml of fluid. The eyes are large and pigmented. The yolk sac is well-developed and the allantoic fluid is clear.

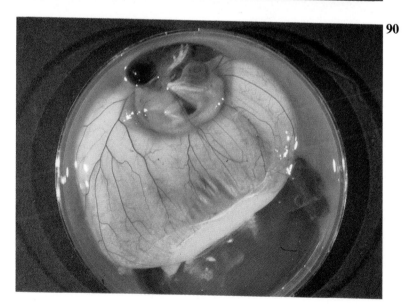

91 Chick embryos at 7, 11 and 16 days. Chick embryos at different stages of development: during the first 10 days cell proliferation takes place, from 10 to 15 days the tissues differentiate and from 15 days to hatching the organs develop to functional state.

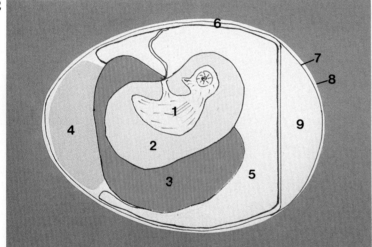

92 Schematic representation of a 10-day-old incubated egg.

1 Embryo
2 Amniotic cavity
3 Yolk sac
4 Egg white
5 Allantoic cavity
6 Chorioallantoic membrane
7 Egg membrane
8 Eggshell
9 Air sac

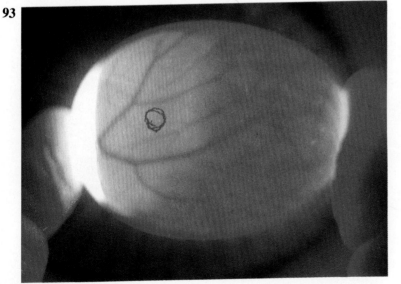

93 Candling the egg (6). Candling for amniotic inoculation at day 10. At this stage, the cavity is well suited for the isolation of influenza viruses from throat washings. The best approach is to hold the egg horizontally during candling and drill a hole just above the amniotic cavity. Inoculation must follow immediately, using the transilluminating lamp.

INOCULATING THE ALLANTOIC CAVITY, HARVESTING ALLANTOIC FLUID

94 A drill with flexible shaft. The egg shell may be cut with a hand held motor tool, but some people have great difficulty using these relatively heavy devices. A motor-driven flexible shaft and a dental handpiece is a good alternative.

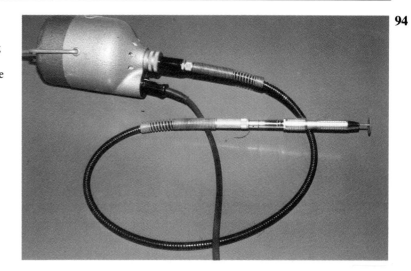

95 Drilling the eggshell. The location of the inoculation hole is marked with pencil during transillumination. The hole is cut with a carborundum disc held in a small hand-drill, which must be reasonably powerful. Care must be taken not to injure the egg membrane.

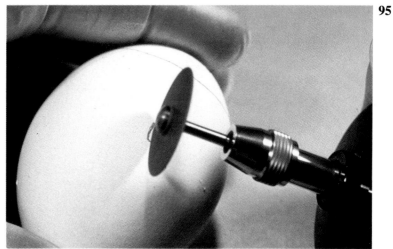

96 Inoculating the allantoic cavity. The inoculum, diluted in a suitable buffer, is injected into the allantoic cavity with a fine needle. This should perforate the egg membrance and the chorioallantoic membrane. Injection has to be slow to compensate for pressure differences, otherwise the fluid will spill from the hole and contaminate the outside of the egg.

97 The sealing mixture. The sealing mixture used to close the eggs after inoculation is prepared by mixing two parts of solid paraffin (melting point ± 54°C) and one part of petroleum jelly (Vaseline). The mixture is kept in a molten state on an electric heating plate (60°C). A short pipette with rubber teat is used to apply the sealing mixture to the eggshell.

98 Sealing the egg. After inoculation the hole is sealed with the paraffin mixture. The sealing compound should not be spilt on the outside of the egg as this inhibits the respiratory function. The sealed eggs must not be turned during further incubation.

99 Opening the egg for harvesting (1). To open the egg the blunt end should be tapped to break the shell. The shell over the air space should be removed by breaking it away piece by piece. All fragments should be considered infected.

100 Opening the egg for harvesting (2). An alternative way of opening an egg is to pierce a hole in the shell covering the air sac with a pair of scissors and cut it open. This is the least time-consuming method but there is more risk of spilling.

101 The air sac of a dead egg. The bottom of the air sac is covered with a whitish, non-transparent shell membrane. When the embryo dies this membrane discolours. These eggs should be discarded if death is due to bacterial infection or lethal trauma.

102 The contents of a 14-day-old egg. Eggs inoculated at 11 days are harvested at 14 days. At this stage the allantoic fluid amounts to 6-9 ml and the amniotic fluid to 1-2 ml. Prior to harvesting fluids containing the virus it is advisable to cool the eggs to 4°C in an icebox for 2-4 hours to kill the embryo. This reduces blood contamination of the allantoic fluid.

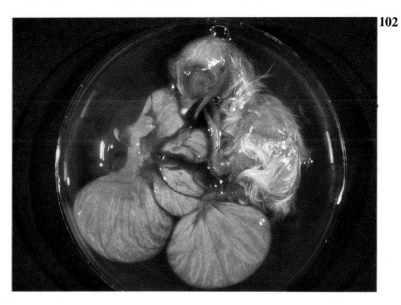

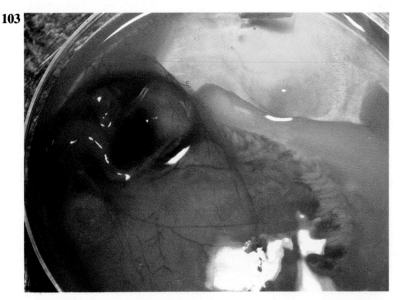

103 The death of the embryo. A number of viruses can kill the embryo. Candling an egg containing a dead embryo shows the disappearance of the typical vein pattern of the chorioallantoic membrane. When harvesting the contents of the egg, the chorioallantoic membrane will not stick to the shell membrane. The embryo is discoloured and shows signs of haemorrhaging.

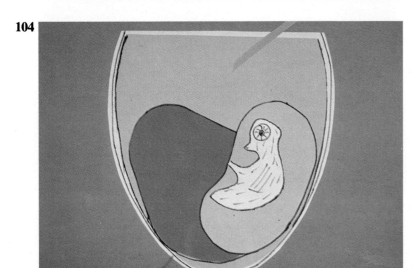

104 Schematical diagram of the harvesting of the allantoic fluid. After cooling, the egg is placed upright. The eggshell covering the air sac is removed. The cooling process prevents bleeding from the vessels.

105 Harvesting the allantoic fluid
The allantoic fluid, floating over the embryo and the yolk sac, is aspirated with a syringe or a pipette: 5 ml or more can be harvested from a 13-day-old egg.

106 Non-agglutinated red blood cells. A suspension of chick erythrocytes in p.b.s.: note the presence of a nucleus which makes these blood cells relatively heavy, resulting in rapid sedimentation. Reactions using chick erythrocytes are complete in 1 hour while those performed with mammalian red blood cells take 2 hours or more to set.

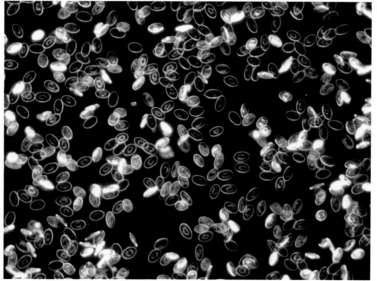

107 Agglutinated chick erythrocytes (influenza). A haemagglutinating virus added to a suspension of red blood cells absorbs on to the cell walls. As the virus is able to attach to more than one cell at a time, bridges are formed between the cells, causing clumping (agglutination) of the cells. These clumps settle at the bottom and form a typical haemagglutinating pattern. The solution must be kept at 4°C to prevent elution of the virus and to stabilise the agglutination.

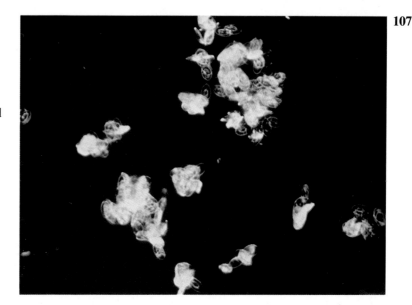

108 Haemagglutination pattern. When amniotic or allantoic fluid is tested for the presence of haemagglutinating virus, equal amounts of egg fluid and red blood cell suspension are mixed in tubes or microplates and put away at 4°C. Agglutination (lower row) is characterized by the formation of a pattern on the bottom of the tube. Non agglutinated blood cells settle down to a small dot (upper row).

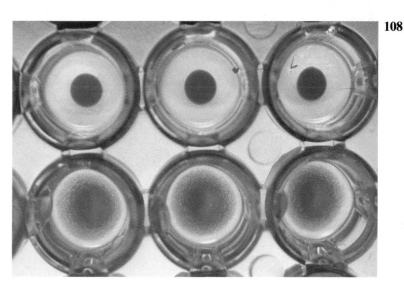

INOCULATION OF AMNIOTIC CAVITY

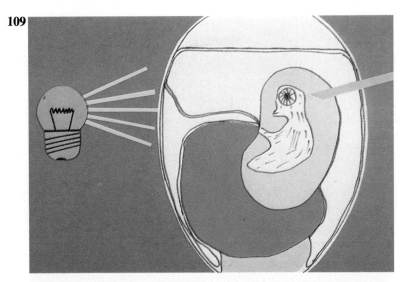

109 Schematic diagram of the inoculation of the amniotic cavity. The needle is placed in the hole drilled just above the amniotic cavity and gently inserted until the amniotic sac moves. Then it is thrust through the amniotic membrane and the fluid is injected slowly. The hole in the shell is closed immediately with sealing material.

110 Inoculating the amniotic cavity. Under visual control 0.1 ml inoculum is injected into the amniotic cavity. A fine needle (gauge 32G x 2.5 cm) is used to minimise the damage to the embryo. The embryo is displaced in the amniotic sac when the needle perforates the amnion.

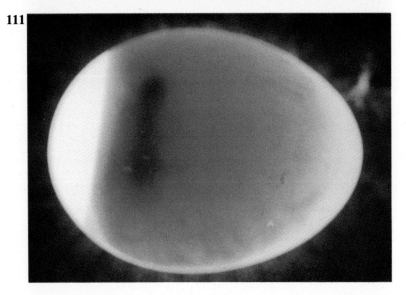

111 Inoculating the amniotic cavity. The inoculated eggs are inspected after 24 hours and those containing dead embryos are discarded. The decaying contents of dead eggs can cause very unpleasant smells in the laboratory.

112 Preparing the eggshell for inoculation.
The shell over the air sac is cut with a rotating carborundum disc. The shell membrane is exposed and made translucent with a few drops of sterile liquid paraffin. Too much paraffin inhibits the respiratory capacity.

113 Inoculating the amniotic cavity. The translucent membrane is pierced and closed forceps are inserted. The forceps are opened and the amniotic membrane is gripped and pulled out. The inoculation is made with a fine needle syringe. This method involves more risk to the embryo than inoculation by candling.

114 Sealing the egg. After inoculation the blunt end of the egg is closed with the original shell cap and molten sealing compound. Alternatively the sealing compound can be applied to the rim of the opening and a cover glass, heated in an open flame, laid on it.

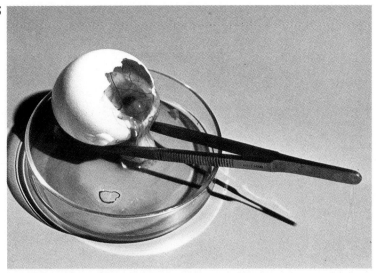

115 Harvesting the amniotic fluid (1). To harvest the fluid the eggshell over the air sac is removed and the membranes cut away. While decanting the allantoic fluid the embryo is manipulated so that it lies just by the opening and the amniotic sac hangs over the rim. The egg is supported by a pair of forceps and the amniotic fluid is harvested with a syringe.

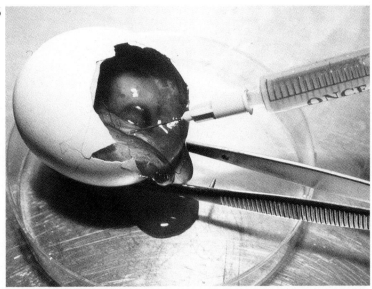

116 Harvesting the amniotic fluid (2). When the egg is 13 days old or more the amount of amniotic fluid is small and hard to harvest. It also contains albumen because at this stage of development the albumen sac is discharged into the amnion. In order to gain a sufficient quantity of fluid for virus investigations the amniotic sac is injected with 2 to 3 ml p.b.s. (phosphate buffered saline). After gentle shaking the fluid is harvested as shown before.

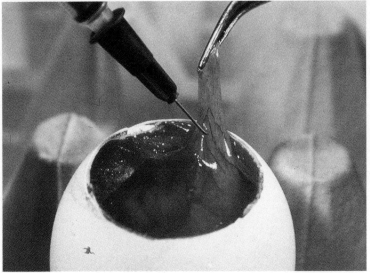

117 Inoculating the yolk sac (1). An alternative method of inoculating the yolk sac is to open up the blunt end of the egg, make the membranes translucent and grasp the yolk sac with a pair of forceps. This method carries a higher risk of trauma and infection, but a larger volume of fluid can be inoculated.

118 Inoculating the yolk sac (2). After candling to determine the position of the embryo, a hole is made in the shell at the centre of the air space. The inoculum is inserted with a long needle and deposited just below the centre of the egg. To ensure the correct position for the needle tip, the plunger is pulled back until yolk can be seen coming up in the syringe.

119 Inoculating the yolk sac (3). After candling and marking the position of the embryo, a hole is cut at 90 degrees in the long side of the egg. It is held horizontally, with the embryo at the top and the yolk sac in the centre. A long needle is used for inoculation.

120 Harvesting the yolk sac. The contents of the egg are poured out into a Petri dish and the yolk sac is separated into another dish. After cutting the yolk sac membrane as much yolk as possible is removed. The yolk sac is transferred to a fresh Petri dish and washed with saline to clear away the residual yolk.

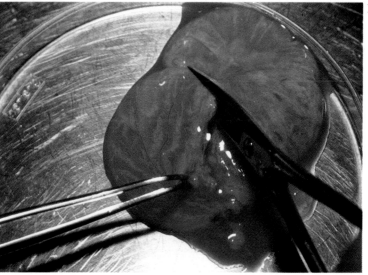

INOCULATION OF CHORIOALLANTOIC MEMBRANE

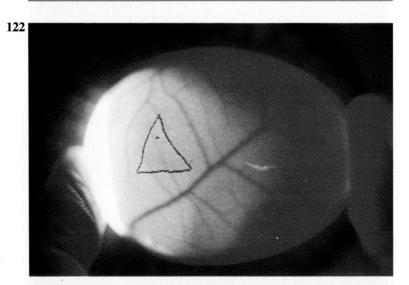

121 Schematical diagram of the inoculation of the chorioallantoic membrane. To inoculate the ectodermal layer of the Chorioallantoic membrane an artificial air space has to be created and the original air sac obliterated. A small equilateral triangle is cut out of the shell and a hole is pierced into the blunt end of the shell, perforating the air sac. After lifting the triangular piece, a small opening is made in the shell membrane and slight suction is applied to the hole in the blunt end. A new air space develops at the top of the horizontally-placed egg and the inoculum is dropped in there.

122 Marking the triangle for the chorioallantoic membrane inoculation. Eggs aged 10-12 days old are candled and a triangle is marked at a site where there are no major blood vessels.

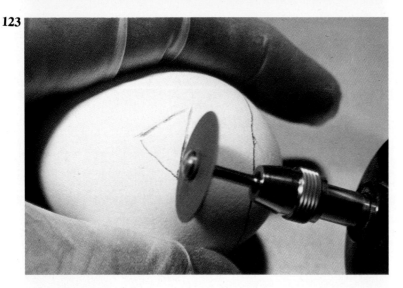

123 Cutting away the triangle for the chorioallantoic membrane inoculation. The shell is cut along the triangle with a rotating carborundum disc. The utmost care has to be taken not to perforate the shell membrane.

124 A badly-cut triangle. When too much pressure is applied to the rotary disc, or the eggshell is unexpectedly thin, the shell membrane is severed, the chorioallantoic membrane is perforated and bleeding occurs. The egg is now unsuitable for inoculation.

125 Making the artificial air space. As the new air space must be made by obliterating the original air sac a hole is pierced in the blunt end to eliminate the original air sac.

126 Eggs prepared for the chorioallantoic membrane inoculation. The triangles are cut but sometimes the shell is not cut all the way through. Therefore the triangle should be gently pressed down to fracture the edges. With some practice the necessary force can be easily judged.

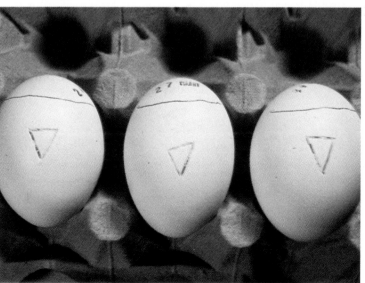

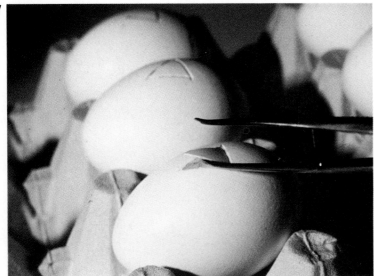

127 Creating the artificial air space. The triangle is lifted out by inserting the tip of a pair of forceps or a blunt needle under the rim. The triangle is put aside for the closure of the egg after inoculation.

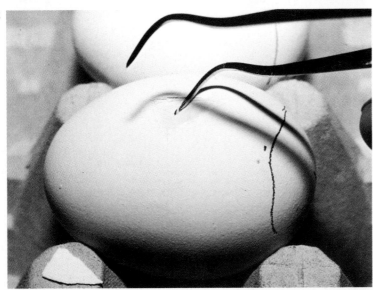

128 Creating the artificial air space. A hole is cut in the shell membrane. Care must be taken not to perforate the chorioallantoic membrane as this renders the egg useless.

129 Creating the artificial air space. The original air sac is removed by perforating the shell membrane in the opening at the blunt end and applying gentle suction with a rubber teat. Forceful suction would damage the shell membrane and pull chorioallantoic membrane through the opening, leading to infection.

130 Inoculating the chorioallantoic membrane. A pipette filled with the inoculum is inserted through the slit in the cell membrane and the contents dripped onto the chorioallantoic membrane. Volumes up to 1 ml can be inoculated. To disperse the inoculum evenly over the membrane the eggs should be rocked gently.

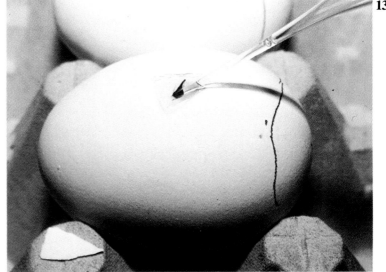

131 Sealing the egg after the chorioallantoic membrane inoculation. The triangle is returned to its position and carefully sealed with paraffin mixture so that the hole is well closed to prevent drying out.

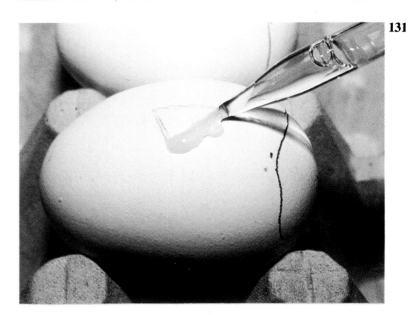

132 Marking the eggs. The type of the virus, the passage number and the date should always be written on the eggshell. A medium pencil should be used because it is atoxic and does not rub off when disinfecting the eggshell.

133 Harvesting the chorioallantoic membrane. To harvest the chorioallantoic membrane the egg is opened at the blunt end. The membranes are cut away and the contents of the egg poured into a Petri dish. Gentle help with tweezers may be necessary to remove the albumen. The chorioallantoic membrane adheres to the eggshell, so it should be pulled out and washed in saline.

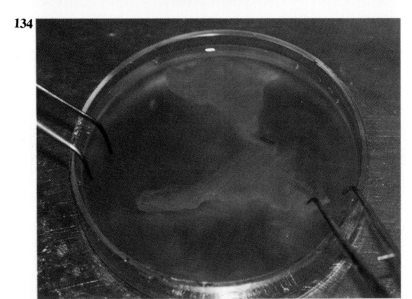

134 Washing the harvested chorioallantoic membrane. The harvested chorioallantoic membrane is washed in two or three changes of p.b.'s. to remove adhering blood, albumen and urates.

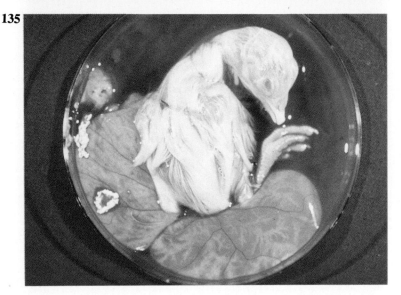

135 The contents of a 16-day-old incubated egg. At this late stage the eggs are used only for harvesting the inoculated chorioallantoic membrane; they are no longer suitable for harvesting the amniotic fluid. The amount of available fluid is small and most of it is trapped between the developing feathers. The allantoic fluid is now heavily contaminated with urates (the whitish substance seen in the left part of the picture).

136 The chorioallantoic membrane at 17 days. As the embryo grows older, urates are deposited in the allantoic cavity. When the chorioallantoic membrane is harvested the urates stick to the membrane and obscure the lesions.

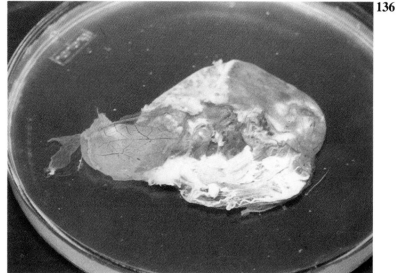

137 Inspecting the harvested chorioallantoic membrane. The harvested chorioallantoic membrane is spread out with two pairs of pincers in a Petri dish filled with sterile saline and set against a dark background. The size (small, medium or large) and the appearance (opaque, whitish or haemorrhagic) of the lesions are noted.

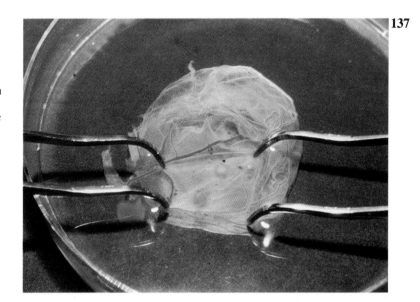

138 Cutting out the virus lesions on the chorioallantoic membrane. After washing the chorioallantoic membrane, the exposed portions carrying the lesions are gripped with forceps and cut out with a small pair of scissors. The collected pieces are transferred to fresh p.b.s. for inspection and further examination.

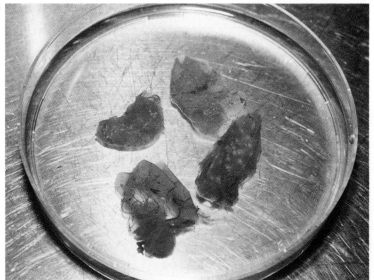

139 Collecting the infected portions of the chorioallantoic membrane. The harvested pieces of the chorioallantoic membrane bearing the lesions are collected in sterile saline. In doubtful cases a bacteriological control examination must be made. The membrane fragments can be stored frozen.

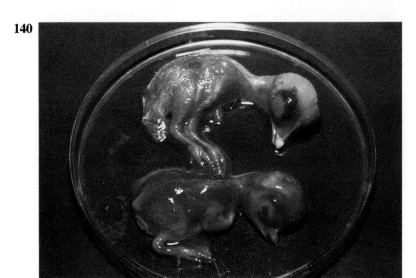

140 Embryo death caused by a virus. A marked retardation in the development and growth of the embryo occurs preceding death. Both embryos shown are 13 days old: the top one is the control, the lower one is infected with vesicular stomatits virus (VSV). It is smaller and shows skin haemorrhages.

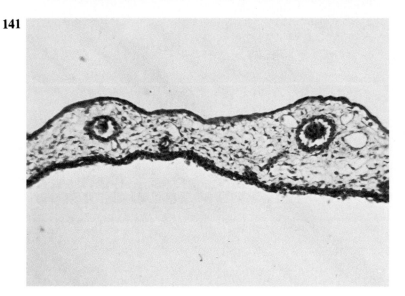

141 The normal chorioallantoic membrane. The normal chorioallantoic membrane is formed by fusion of the chorion and the allantois. The membrane is covered externally by chorionic ectoderm and internally by allantoic entoderm. The mesoderm situated between these layers contains numerous blood vessels. The chorioallantoic membrane, adhering to the shell membrane, is the main respiratory organ of the embryo.

142 The chorioallantoic membrane inoculated with influenza virus. The inoculation of influenza virus into the allantoic cavity does not cause morphological changes in the chorioallantoic membrane.

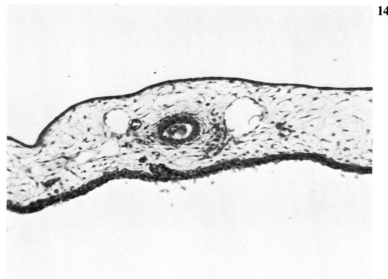

143 Non-specific lesions of the chorioallantoic membrane. Suspension or extracts dropped on to the ectoderm of the chorioallantoic membrane can cause non-specific proliferation and induration of the membrane. These simulate the effects of virus infections but are actually caused by irritation.

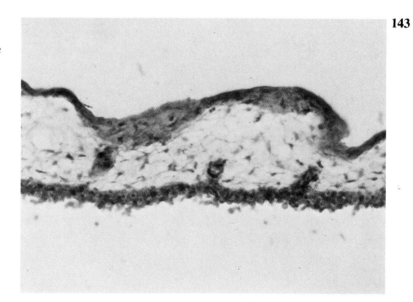

144 Lesions of the chorioallantoic membrane caused by C.E.E. virus. The chorioallantoic membrane of a 10-day-old incubated egg was inoculated with a diluted suspension of mouse brain infected with Central European Encephalitis (C.E.E.) virus and incubated for another 3 days. The alterations caused by the proliferation of the tick-borne virus consisted of small whitish dotted lesions of approximately 0.3 mm.

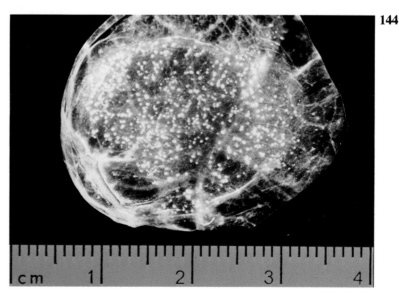

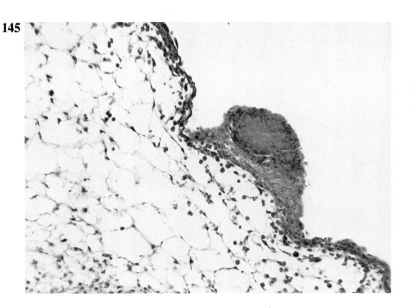

145 The histology of a C.E.E. lesion on the chorioallantoic membrane. The lesion of the C.E.E. virus on the chorioallantoic membrane is circumscribed and consists of proliferating ectodermal cells with central necrosis. Oedema and cellular infiltration are practically absent from the mesoderm.

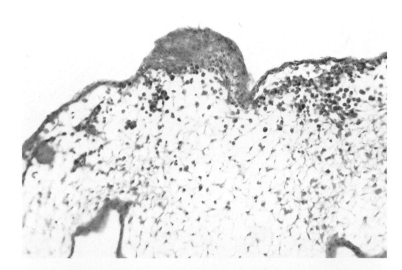

146 The histology of a louping-ill virus lesion on the chorioallantoic membrane. In this lesion degeneration of the proliferating cells is absent and the mesoderm shows infiltration. The lesions retrogress early but the embryo remains subacutely infected and dies if incubation is extended.

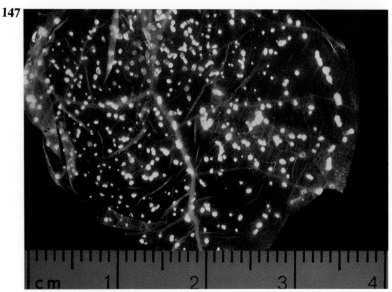

147 Herpes virus hominis Type 1 lesions on the chorioallantoic membrane on day 3. Typical appearance of the chorioallantoic membrane of a 13-day-old egg on the third day of incubation after inoculation of vesicle fluid from a lesion of the upper lip. The numerous lesions are small, white and circumscribed and the mesoderm is not oedematous.

148 The histological appearance of a 3-day-old herpes virus hominis Type 1 lesion. A section of choriallantoic membrane on the third day of inoculation with herpes virus hominis Type 1. The lesion shows crater-shaped necrosis and slight cellular infiltration in the mesoderm.

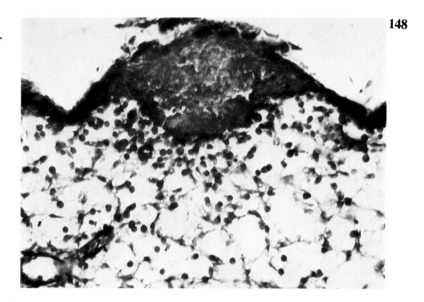

149 Herpes virus hominis Type 1 lesions on the chorioallantoic membrane on day 7. Compared with the appearance of the membrane on the third day the lesions of the Type 1 virus enlarge only slightly if the incubation period is extended from 3 to 7 days. In contrast the lesions of the Type 2 virus enlarge considerably during prolonged incubation.

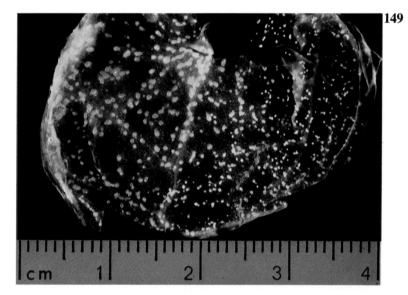

150 Herpes virus hominis Type 2 lesions on the chorioallantoic membrane on day 3. On the third day of incubation the lesions of the Type 2 virus resemble closely those of the Type 1 virus and a differential diagnosis cannot be made on the macroscopic appearance.

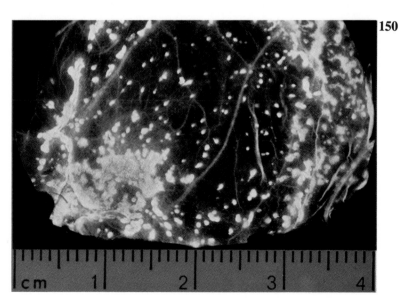

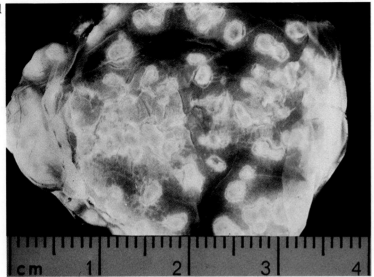

151 Herpes virus hominis Type 2 lesions on the chorioallantoic membrane on day 7. During prolonged incubation the lesions of the Type 2 virus enlarge considerably and the chorioallantoic membrane becomes oedematous. The example shown is so characteristic that the diagnosis of Type 2 virus is certain.

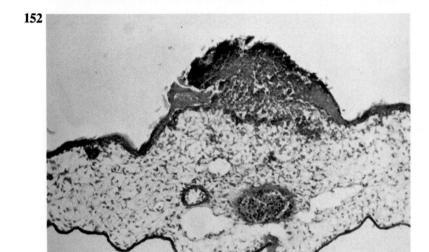

152 Histological appearance of a 7-day-old herpes virus hominis Type 2 lesion. At 7 days the Type 2 lesion shows a rim of proliferating ectodermal cells and necrosis of the central part with cellular infiltration. The mesoderm is oedematous and infiltrated with cells.

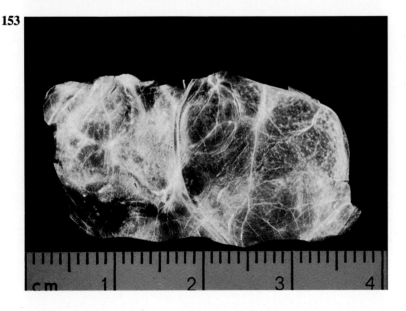

153 Lymphocytic choriomeningitis lesions on the chorioallantoic membrane on day 3. In most cases the lymphocytic choriomeningitis virus produces trivial lesions on the chorioallantoic membrane. The lesions shown are not pathognomonic.

154 Vaccinia virus lesions on the chorioallantoic membrane on day 3. The lesions of vaccinia virus show a marked profileration of the ectoderm with central necrosis giving a white appearance. A number of small secondary lesions are present on the third day. When a small piece of membrane with lesions is cut off and rubbed onto a slide the virus particles can be stained and observed by high power light microscopy.

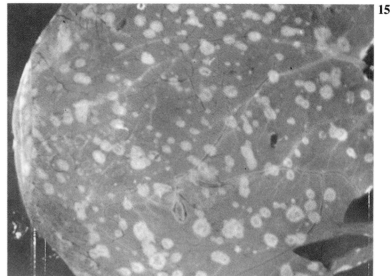

155 Section of cowpox lesion on the chorioallantoic membrane. The pox viruses can produce cytoplasmic inclusion bodies in the infected cells. In the case of cowpox these bodies are very prominent and pathognomonic. Lendrum staining has turned the typical inclusion bodies red.

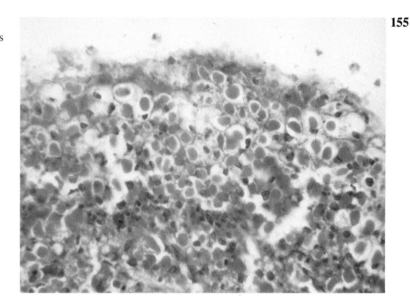

156 Removing a pox virus lesion for staining. A piece of the chorioallantoic membrane containing lesions is removed and rubbed on a slide. The cells are thus ruptured and the elementary bodies are spread out. The slide can be stained by the Morosow method.

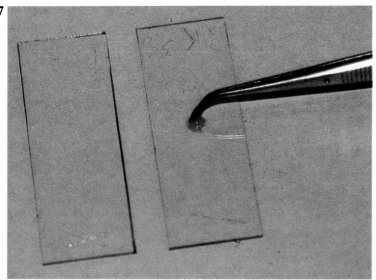

157 Making a smear of a pox lesion. The elementary bodies of the pox viruses shown in the lesions of the inoculated chorioallantoic membrane. To make a smear the lesion is cut out and the tissue fragment is held between pincers, as shown, and rubbed on a clean glass slide. As the preparation is to be inspected by an oil immersion lens, care must be taken to place the smear in the middle of the slide (as shown in the next illustration) to prevent a spilling of the microscope oil.

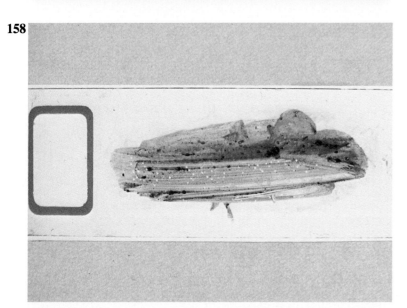

158 A typical smear of a chorioallantoic membrane lesion of a pox virus. The slide is stained with the Morosow stain, a silver impregnation, which reveals the presence of elementary bodies.

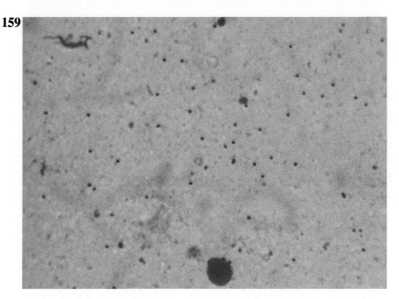

159 Elementary bodies in a silver-stained smear. The thin spots of the Morosow stained slide are inspected with high magnification. The condensor system must be carefully adjusted. The virus particles are shown as dark, uniformly formed points, which are alike for the different types of pox viruses. So the only information obtainable from a Morosow preparation is: pox virus or not pox virus.

Experimental animals

The use of experimental animals has always been important in virus work. Using them as a means of isolating or growing of viruses is only one of the reasons for using animals. From an ethical standpoint as well as for financial reasons the use of experimental animals must be restricted to the absolutely necessary. Scientific indications for animal experiments occur in the following circumstances and cases.

The virus will not replicate in cell cultures or embryonated eggs or the virus grows better in experimental animals.

It is important to study the pathological processes of the disease caused by the virus; experimental animals can be of great help, as is also the case in epidemiological investigations.

Animals can be used for attenuation of human viruses, for example yellow fever and vaccinia virus.

Safety and antigenicity testing of vaccines is first done in animals before a vaccine can be licensed.

Animals are indispensable for testing antiviral chemotherapeutics.

The production of polyclonal and monoclonal antibodies must be done in animals. Human sera are of little use in diagnostic and experimental virology.

Animals are a source of organs for cellcultures.

Animals are a source of erythrocytes used in many reactions.

A number of viruses will not replicate in cellcultures or embryonated eggs but will do so in certain experimental animals. In diagnostic virology this is the case with most Coxsackie A viruses. Practically all of the so called arboviruses can be readily isolated from blood and mosquito pools by inoculation into the mouse. Rabies virus cannot be diagnosed in tissue culture but is highly pathogenic for mice. The suckling mouse is ideal for isolating Foot and Mouth disease virus.

In some cases the use of experimental animals will speed up the diagnosis and will give important clues as to the nature of the virus involved. This is, for example, the case in Coxsackie B infections, the virus of which readily grows in cell cultures but with a cytopathic effect which closely resembles that of other enteroviruses. In suckling mice typical signs and lesions will yield the definitive diagnosis. Sometimes the experimental animal is more sensitive to the infection than the other systems and smaller amounts of virus can be detected as can be the case with some herpes simplex strains when inoculated intracerebrally in suckling mice.

As a number of viruses replicates better in animals higher titres can be attained than in any other system for the production of antigens. This is the case for the so-called arboviruses and for most Coxsackie A strains. Even if the virus grows better in cell cultures it can be an advantage to inoculate animals. An example is poliomyelitis virus which when inoculated in rhesus monkeys will give rise to typical pathological lesions.

The use of experimental animals can have disadvantages. Animals harbour latent viruses which will contaminate virus strains that are maintained by passage in these animals. There can be variations in susceptibility for viruses due to differences in race, sex and age. When animals are infected they can spread the virus by urine, faeces, saliva etc. to other animals or to people in their surroundings. Animal rooms must be designed to give optimal conditions for the animals and to minimize the hazards of infection. Strict rules must be followed.

PRODUCTION OF ANTISERA

Large amounts of antibodies can be produced by immunization of experimental animals. The resulting specific antibodies are of polyclonal origin and are directed to a broad range of antigens of the virus involved, an advantage when the sera are used to type viruses isolated from patients. Sometimes fine differences in strains have to be detected and a smaller range of specificity is needed, in that case monoclonal antibodies are to be used.

To obtain animal antisera of the best possible quality a number of rules have to be followed and precautions taken. Use young healthy animals, when available s.p.f. (specified pathogen free), eventually gnotobiotic. A sufficiently large pre-immunization serum sample must be taken in order to control the eventual presence of pre-existing antibodies which might interfere with the tests done with the immune serum.

Two types of immunization procedures can be distinguished on the basis of whether the animal is susceptible to the virus or not. If the animal can be infected with the virus, antibodies are naturally induced to the antigens of the virus only. Sera of this type can be used for complement-fixation, neutralization, haemagglutination inhibition and so on. The immunizing procedure varies. Some viruses can be instilled intranasally as is the case with influenza in ferrets, others can be applied to the scarified skin as can be done with herpes simplex in rabbits.

It may be necessary to infect the animal by injection. In that case it is important that the antigen does not contain non virion proteins, which are foreign for the animal, in order to prevent antibody formation against these foreign proteins. For example, when immunizing a rabbit against rubella virus the virus must be grown in a primary rabbit kidney cell culture with maintenance medium containing rabbit serum. The immunization procedure can be done by 5 or 6 weekly intravenous injections without risk of anaphylactic shock or induction of unwanted antibodies. Coxsackie antisera are, for the same reasons, made by injecting adult mice intraperitoneally with high titred suckling mouse derived virus. Three to five injections every two weeks will do. The titre can be controlled during the immunization period by taking bloodsamples using the method of the orbital puncture.

When animals are not susceptible to the virus to be used immunization must be done by injecting considerable amounts of antigen. This brings with it the risk of inducing antibodies against impurities in the antigen preparation. Sera prepared this way can be cytotoxic in neutralization tests, will give false positive complementfixation tests, or give rise to much background fluorescence in I.F. techniques. To overcome these disadvantages the antigen must be purified as highly as possible. Mostly the antigen is mixed with an equal volume of Freund's complete adjuvant. The mixture can be given subcutaneously or intramuscularly. Do not overimmunize as this will boost the titre against traces of impurities. The specificity of sera can be enhanced by suitable adsorbtion procedures.

The height of the titre can be controlled regularly by taking bloodsamples. From small animals by orbital puncture, from rabbits from the marginal vein of the ear. When sufficiently high titres are reached the animals are bled. Small animals are bled to death under anaesthesia. The guinea pig is bled by cutting the carotis. The mouse is given intraperitoneal anaesthesia and bleeding is done by cutting the axillar arteries. Larger animals can be tapped several times. The rabbit is bled by puncturing the artery of the ear with a 21J½ needle and about one third of the circulating volume can be taken. This procedure can be repeated several times.

Serum is made by the following procedure: the blood is kept at room temperature for 15-20 minutes in order to clot, then the clot is loosened from the wall and the tubes are placed in the incubator for half an hour to enhance retraction of the clot and then put in the refrigerator at 4°C overnight. The following morning the serum is separated by centrifugation. The serum from all the tubes from the same animal is pooled and titrated. Serum keeps well for many years in the deep freeze at $-20\text{-}30°C$, but freezing and thawing causes deterioration. The best way to preserve antisera is filling 0.5-1.0 ml ampoules and subsequent lyophilization.

PRACTICAL USES FOR ANTISERA

Antisera are used for three purposes.

1 *As control sera in serological reactions.* These sera must be sufficiently specific to be used with different batches of antigens. Human sera, when properly selected, can be employed as controls for many routine serological tests. For example for cytomegalo c.f.t. (complement fixation test), rubella h.a.i. (haem agglutination inhibition) and haemolysis in jel (h.i.g.). Select the sera on the basis of the height of the titre, the presence or absence of IgM, the absence of reactions to control antigens and crossreactions to other antigens. Animal antisera can be used when they are made according to the specifications of the reaction for which they are used. When sera are to be used as control for the complement fixation test the animal must be immunized by infection. Immunization by injection of antigens always gives rise to non-specific titres in the complement fixation test. Animal sera for the haemagglutination inhibition test can be made by immunizing the animals by the parenteral route as antibodies to tissue and serum components will not be adverse to the reaction.

2 *As sera for the identification of viruses.* Human sera cannot be used as they contain a great number of different neutralizing antibodies, which makes a reliable virus identification possible. Animal sera for the neutralization test, for the haemagglutination inhibition test and for the heaemadsorbtion test can be made by parenteral immunization of animals. Antibodies to non-virion antigens will not influence the identification reaction.

3 *As sera for the detection of virus.* When animal sera are used for the detection of virus in clinical specimens by the immunofluorescence method or immuno-enzyme techniques the quality must be such that the virus antigens are clearly stained and the background not. Sera are prepared by injecting animals or using highly purified antigens for the parenteral immunization. In most cases sera must be absorbed by human tissue to make them more specific. Sera for the capture technique with ELISA and RIA must comply to the same high qualities.

SUCKLING MICE IN DIAGNOSTIC VIROLOGY

160 Breeding mice (1). To breed albino mice, one 4-month-old male is brought together with 10-12-week-old females. Fertilization takes about a week, so that not all litters are born on the same date.

161 Breeding mice (2). The pregnant mice are separated. In this picture a stainless steel breeding system is shown in which the mice are kept on sawdust. An automatic drinking system is provided. A rack at the front holds food pellets sufficient for a week.

162 Breeding mice (3). The inside of a tray. The litter is kept together during 3-4 weeks when the mice weigh 15–18 gm. They are then weaned, sexed, and put in larger trays until needed for tests or breeding.

163 Suckling mice. A nest of suckling mice which must be inoculated with contagious material must be kept in a separate cage. This one is made of macrolon, which is steam-sterilizable, with a stainless steel lid with a pellet rack and drinking bottle.

164 Suckling mice. A litter of 8 three-day-old suckling mice. The mice are healthy and can be handled without damage. Depending on the virus to be inoculated the mice are used new-born for Coxsackie A, up to 2 days for Coxsackie B, up to 3 days for herpesvirus.

165 Suckling mice. A healthy suckling mouse is well cared for by the mother. The stomach is filled with milk and the bladder is hardly visible. When suckling mice are ill they do not drink and the bladder is overfilled due to the fact that the mother will not lick to assist micturition.

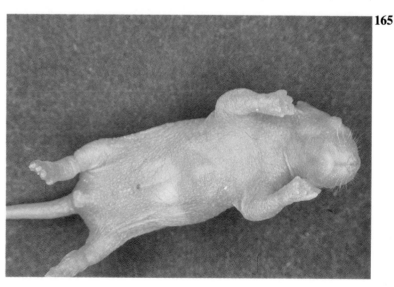

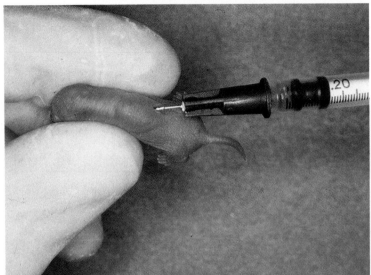

166 Inoculating a suckling mouse. The mouse is held between thumb and finger. Inoculation is mostly done by the subcutaneous route at the backside. The intracerebral and the intraperitoneal route are also used; sometimes combined routes are used.

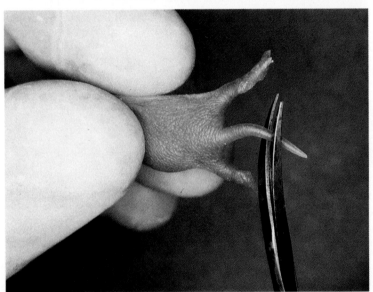

167 Marking suckling mice (1). A standard litter of mice is eight. For most titrations half a litter will do for each dilution. Four mice will have normal tails, the others are marked by cutting off the tip of the tail.

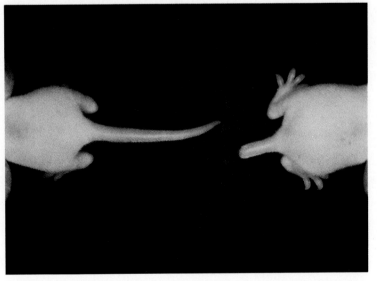

168 Marking suckling mice (2). At ten days the difference in length is obvious and makes sorting easy.

169 Dissecting the suckling mouse (1). The mouse is killed by cooling at 4°C for half an hour. Start the dissection by pulling off the skin with two pincers. Break the skin at the backside and pull the skin to the front and the back.

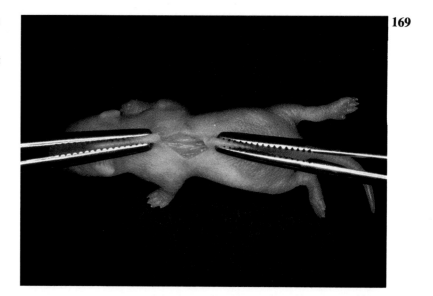

170 Dissecting the suckling mouse (2). Pull the skin over the front part and cut off the fore paws to pull the skin over the head. Remove the hindpart of the skin and cut off the tail at the base and the hind paws.

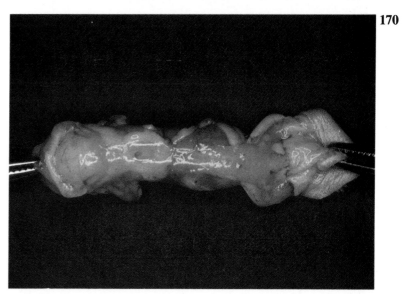

171 Dissecting the suckling mouse (3). When the carcase is needed, for example in Coxsackie A and B, the abdominal organs must be removed. Tear off the abdominal wall and use a tweezer to tear out the packet of intestinal organs.

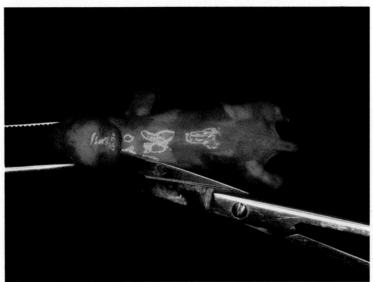

172 Dissecting the suckling mouse (4). After skinning, the organs can be removed. If there is an encephalitis the brain is harvested by cutting open the calvarium and lifting out the brain.

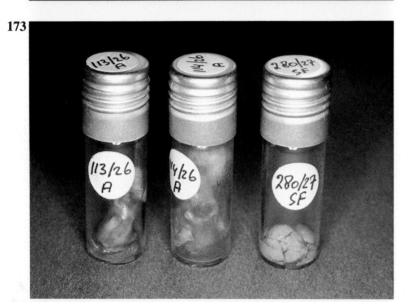

173 Storing small samples in the deep freeze cabinet. Flat-bottomed glass containers with screw caps containing a sealing inlay are most convenient. To prevent cross contamination both caps and tubes are labelled. The tubes can be stored in boxes or special racks.

174 Making a suspension of mouse tissue. Virus is extracted from infected mice carcasses by grinding with sterile sand and making a suspension in saline. After centrifugation the supernatant is used as a virus source.

INOCULATION OF SUCKLING MICE

175 Coxsackie A. Most of the Coxsackie A viruses must be isolated by inoculating new-born mice. In a few days a marked necrosis of the striated muscles develops, which causes flaccid paralysis. Pinching the tail will not cause strong motility as is the case in healthy mice.

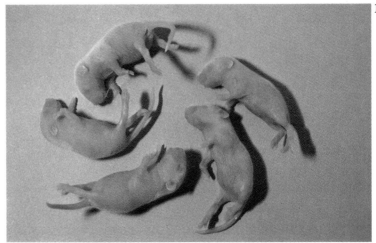

176 Macroscopic appearance of Coxsackie A. Coxsackie A causes necrosis of the skeletal muscles which manifests itself as white discolorations. In the illustration this can be seen in the long back muscles and the muscles of the thigh. In most cases the intercostal muscles are affected.

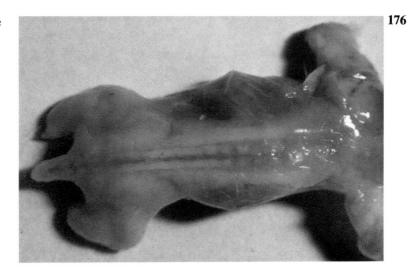

177 Signs of illness in a suckling mouse. Suckling mice stop drinking milk when they become ill. The mother mouse will then no longer take care of the baby and help to empty the bladder. At dissection the overfilled bladder and the empty stomach are typical signs of illness.

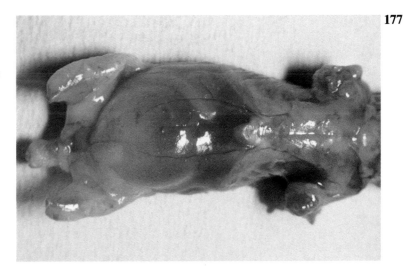

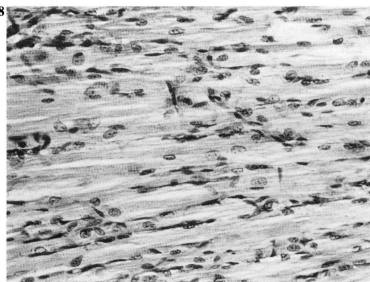

178 Normal striated muscle. The striated skeletal muscle of the uninfected suckling mouse showing all the histological characteristics of healthy muscular tissue.

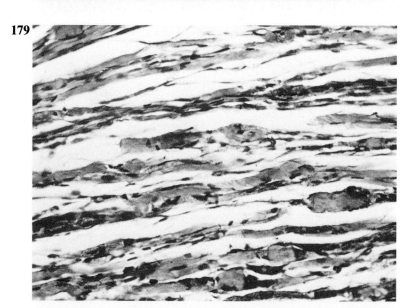

179 Myositis in Coxsackie A infection. The muscle lesion of a suckling mouse infected with Coxsackie A virus is characterized macroscopically by its whitish colour. Histologically, considerable damage can be observed: oedema, waxy degeneration and coagulation necrosis with loss of the cross-striations.

180 Coxsackie B. Two suckling mice, who were inoculated with a faecal extract of a patient with Bornholms disease. They have paralysis due to infection of the central nervous system. There is also a necrosis of the brown fat and sometimes a pancreatitis.

181 Macroscopic appearance of Coxsackie B. The most typical appearance is the greyish-white discoloration of the interscapular brown fat, with focal opacities. Sometimes there is a slight involvement of the skeletal muscles, but never to the extent seen in Coxsackie A.

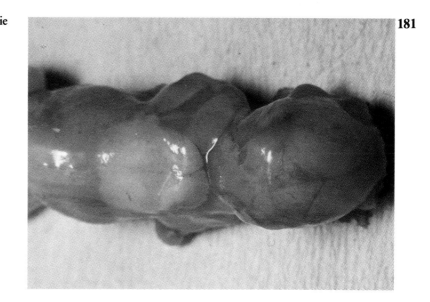

182 Histological section of a Coxsackie B mouse. As soon as symptoms of Coxsackie B infection occur one mouse is used to make the histological diagnosis. The carcase is fixed in formalin 10% and processed for histological sections. The C.N.S. mostly will show necrosis of neurons. Lesions in the interscapular fat and in the pancreas can be detected.

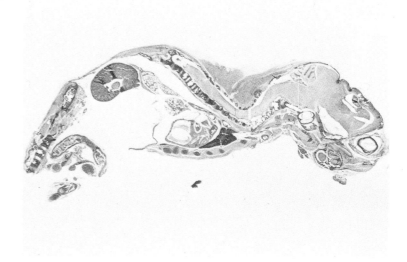

183 Fat necrosis in Coxsackie B. The panniculitis consists of focal necrosis caused by the death of young fat cells, followed by subsequent healing, regeneration and calcium deposition.

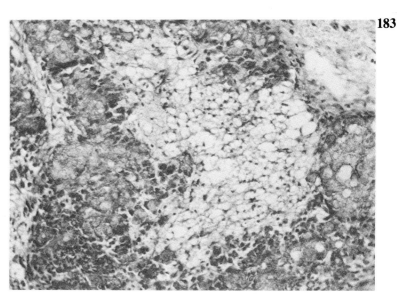

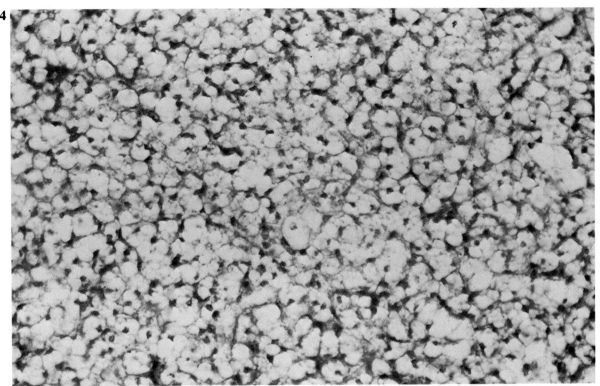

184 Normal brown adipose tissue. The interscapular fat pad consists of brown adipose tissue, more dense than regular fat.

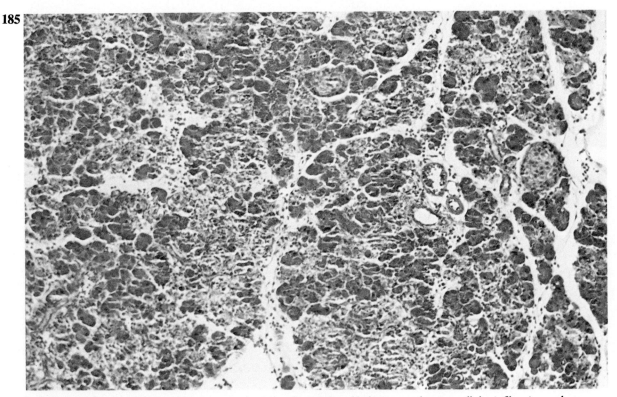

185 Pancreatitis in Coxsackie B. The pancreas of a Coxsackie B infected baby mouse showing cellular infiltration and degenerated acinar cells. The ducts are spared. The islets of Langerhans are not involved.

MARKING MICE

186 Ear-marking individual mice (1). The nine possibilities of ear-marking mice with a punch.

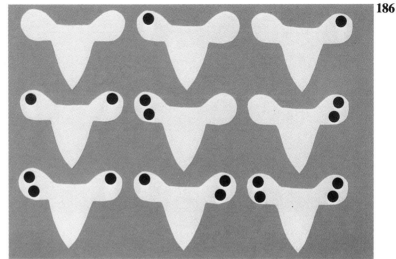

187 Ear-marking individual mice (2). A more permanent marking can be done by punching holes in the ears of the mouse. The animal is fixed by pressing down on a cork plate and the required marks are punched out without anaesthesia. The punch shown was custom made.

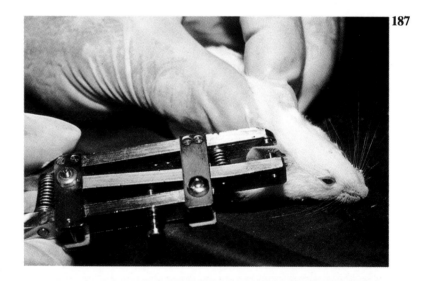

188 Ear-marking individual mice (3). The punched hole will never close again, which makes the marking system permanent.

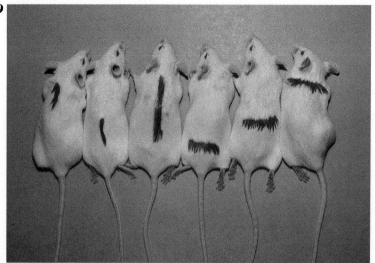

189 Marking mice with dye. The individuals from a group of mice can be marked temporarily with dye, for example with indelibile felt-tipped pens. Different colours can be used to distinguish the groups.

190 Inhalation narcosis. Inhalation narcosis is done by placing the mouse in a jar in which tissue paper or cotton wool is impregnated with ether. Observe the mouse carefully as a prolonged exposure may be fatal. The narcosis is of short duration and given for minor operations like orbital bleeding or intra-nasal infection.

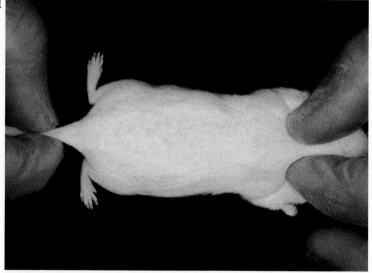

191 Killing the mouse by cervical dislocation. The quickest way to kill a mouse is by cervical dislocation. Put thumb and finger behind the head, press firmly down and pull the base of the tail with the other hand. This method is fit for all occasions except those in which blood is needed, because of the extensive bleeding in the tissues of the neck.

192 Dissecting the mouse (1). After killing the mouse, the skin is wetted with diluted alcohol, not to disinfect but to keep the hair from being spread over the working space.

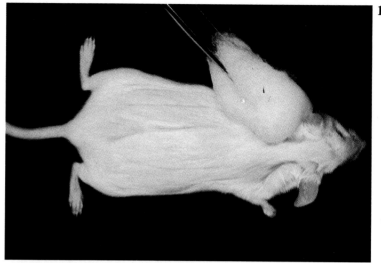

193 Dissecting the mouse (2). Cut the wetted skin in the middle of the back.

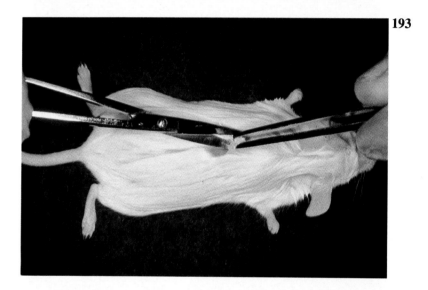

194 Dissecting the mouse (3). With the help of two surgical tweezers the skin is pulled backwards and forwards.

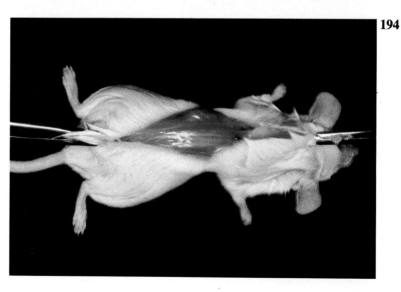

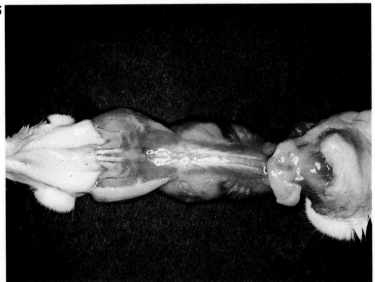

195 Dissecting the mouse (4). Here the body of the mouse is completely denuded. When this is done carefully the surface is sterile and does not need any disinfection for further work. Internal organs like the spleen and the kidney can be clearly seen.

196 Dissecting the mouse (5). To remove the brain the wetted skin is cut away over the head. The top of the skull can be cut away with a pair of small scissors. The brain is lifted out by placing the scissors in a slightly opened position under the brain.

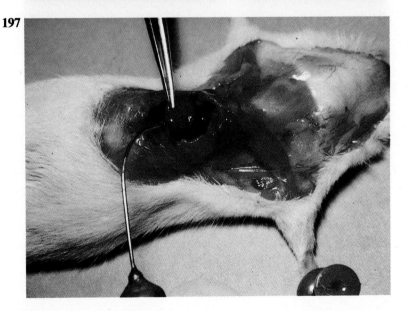

197 Perfusion of the mouse. For perfect fixation of organs, perfusion with formaline is used. This mouse is given a deep narcosis and is pinned down. The skin above the thorax is cut away, the sternum is removed, a cut is made in the right atrium and in the left ventricle. A round tipped needle is brought via the cut in the right ventricle wall into the aorta.

198 Perfusion of the mouse, diagram. The drawing shows the openings made in the right atrium and the left ventricle. The needle is inserted via the ventricle into the aorta. A 20 ml syringe is used to press the formaline solution by a tube attached to the needle through the circulation of the mouse. The blood and the superfluous formaline leaves the body through the cut in the atrium.

1 Cut in the left ventricle
2 Cut in the right antrium
3 Blunt or round tipped needle for perfusion
4 Needle tip through the aortic valve

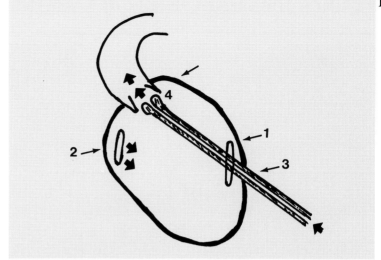

199 A perfused mouse. Fixing by perfusion is so thorough that the mouse is completely stiff. The organs which are needed for histology can now be taken out.

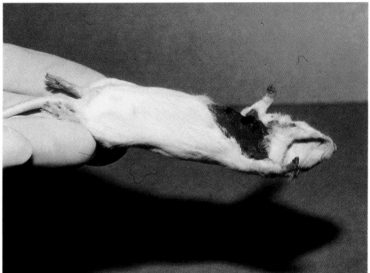

HANDLING LABORATORY ANIMALS

200 Bleeding mice (1). Blood samples of 0.1 to 0.5 ml can be collected from mice by puncture of the orbital plexus with a finely drawn pasteur pipette. The mouse is given a light ether anaesthesia. Vision will not be impaired. Samples can be taken at least once a week.

201 Bleeding mice (2). When the mouse is sacrificed for blood collection axillar bleeding is a useful way of harvesting the blood. The mouse is put under anaesthesia and pinned to a solid surface. The skin above the axillar region is opened and the arteries cut. The blood collects in the wound and can be collected with a pipette.

202 Bleeding mice (3). Small amounts of blood can be collected by warming the tail to stimulate the blood flow and cutting off the tip of the tail. Do not massage the tail, as the blood will be contaminated by tissue fluids.

203 Intramuscular injection. Intramuscular injection is done in the thigh. Anaesthesia is unnecessary. Fix the mouse as shown, grabbing the skin of the back between thumb and fingers. Always disinfect the skin with iodine to prevent abscesses.

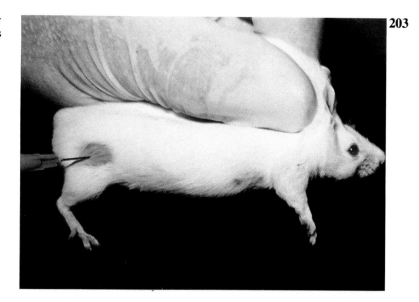

204 Intracerebral injection. Fix the mouse by firmly grabbing the skin of the neck and pressing down the head on the table. Inject 0.01 to 0.02 ml slowly.

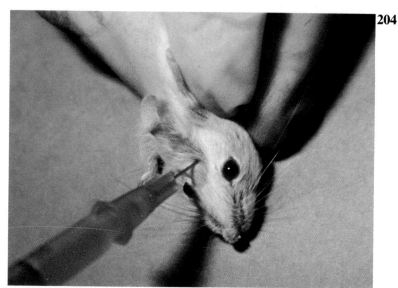

205 Intravenous injection. Fix the mouse in a plastic holder with a slit to pull the tail through. Rub the tail with xylene to cause hyperaemia; the vein will show clearly.

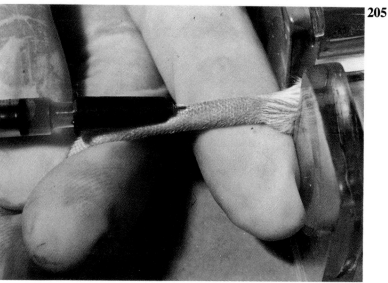

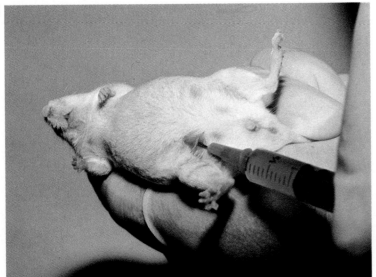

206 Intraperitoneal injection. Fix the mouse by grabbing the skin of the neck and the back in one hand. Disinfect the skin of the injection site. Hold the mouse with the head somewhat downward to move the intestines. The needle is pushed in slowly to prevent perforations.

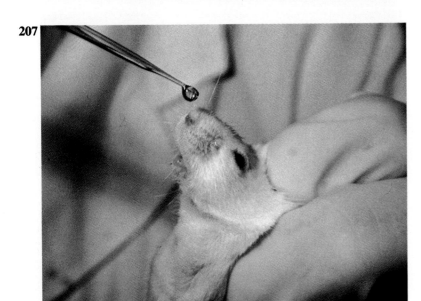

207 Intranasal infection. For intranasal infection ether narcosis is given. Hold the mouse upright and apply a drop of fluid at the moment of inspiration. As aerosols are formed care must be taken not to infect oneself.

208 Isolator. Infected animals which must be kept apart can be kept and handled in an isolator in which the incoming and exhaust air is filtered. All handling is done by gloves built into the wall.

209 Encephalitis in the mouse. During the first phase of an encephalitis the c.n.s. is excited. This mouse, intracerebrally infected with Japanese B encephalitis virus, shows the typical cramps of the excitation.

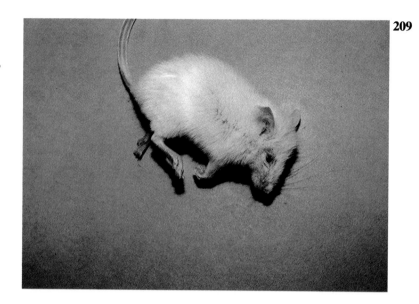

210 Paralysis. A fulminant encephalitis always leads to paralysis. The legs lack power and can hardly be moved. In a few hours this mouse died of his Japanese B virus infection.

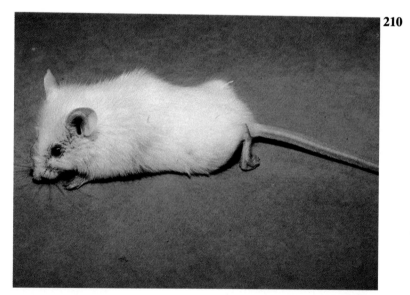

211 Central European Encephalitis. When the virus of the Central European Encephalitis (C.E.E.) virus is inoculated intracerebrally, the first signs of the disease are impairment of the movement of the hind legs. The photograph shows the typical posture of such legs.

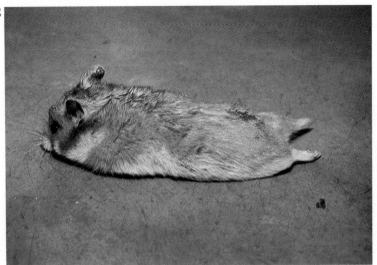

212 Herpesvirus. Hamsters are extremely sensitive to peripheral infection with herpesvirus. This hamster was infected on the lower back skin. The virus travelled via the sensible nerves to the spinal cord and induced paralysis of both hind legs.

213 SV40 virus. Tumours can be induced in hamsters by infecting them with viruses not giving rise to tumours in the original host. The hamster shown here is bearing an enormous subcutaneous tumour induced by SV40 virus to which the animal was exposed in the weaning period.

214 Reo virus. This nude mouse was infected just after birth with reo virus type II. The virus interfered with his normal development and resulted in this runt mouse. Reo virus infections are extremely detrimental to mouse breeding colonies.

215 Bleeding guinea pigs. Total bleeding of guinea pigs is done by bringing the animal under anaesthesia and cutting the carotid artery with a sharp knife. The blood is collected in a centrifuge tube and allowed to clot.

The technique is used when making complement and at the end of an immunization procedure.

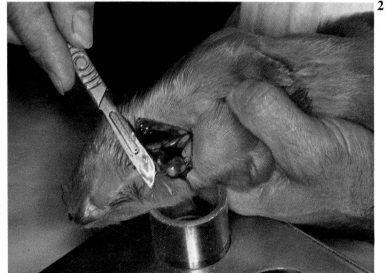

216 Heart puncture in the guinea pig. Carefully palpate the thorax to locate the heart. Use a needle on a syringe to puncture the left ventricle. Some experience is needed to know how deep the needle must go and how quickly blood can be drawn into the syringe. When the blood is needed for erythrocytes, heparin or sodium citrate is used in the syringe to prevent clotting.

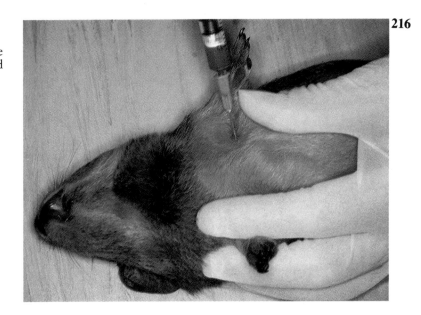

217 Orbital puncture. Orbital puncture is done with a capillary tube under general anaesthesia. The blood flows into the centrifuge tube. This is the way to take blood samples during an immunization.

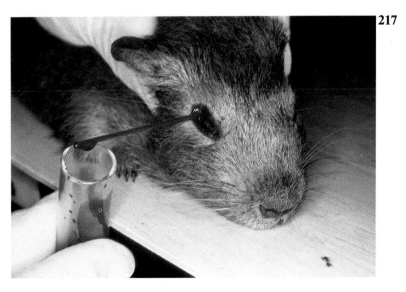

218 Ear-marking rabbits. Rabbits can be marked permanently by tattooing the ear with a number. When many rabbits are used it is more practical to use metal earclips with a number. Apply the clip to the frontal rim of the ear as the vein runs along the distal rim.

219 Metal ear-markers. These metal ear-markers are closed around the rim of the ear. Do not use pliers to press the two parts together as this will lead to local necrosis and the loss of the clip.

220 A rabbit box. Rabbit boxes are used when blood is taken, an injection is given, and in all cases when immobilization is important. The rabbit is fixed in the hole by pressing its backside forward with an adjustable panel.

221 Bleeding the rabbit (1). Rabbits can be bled from the vein and from the artery of the ear. To stimulate the blood flow the ear is rubbed with a cotton swab wetted with xylene.

222 Bleeding the rabbit (2). The vein is cut and the blood collected in a centrifuge tube. When the room is warm and the ear is well rubbed, 10 to 12 ml of blood can be collected.

223 Bleeding the rabbit (3). When larger amounts of blood are needed the arterial puncture is the best way to collect up to 30 ml of blood. A healthy rabbit can be bled this way weekly, greatly expanding the harvest of immune serum.

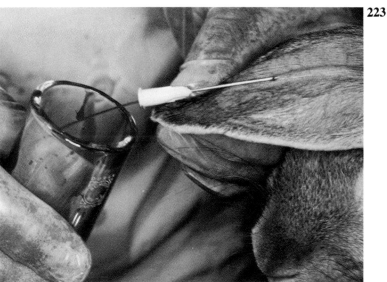

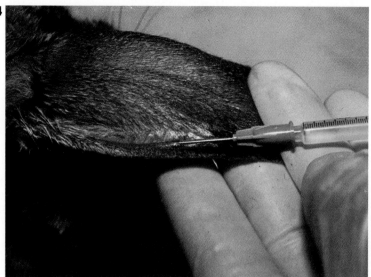

224 Intravenous injection. Intravenous injection in a rabbit is done in the marginal vein of the ear. Always pull the plunger of the syringe to let some blood appear in the fluid to be sure that the needle is in the vein. Otherwise local inflammation and necrosis can occur.

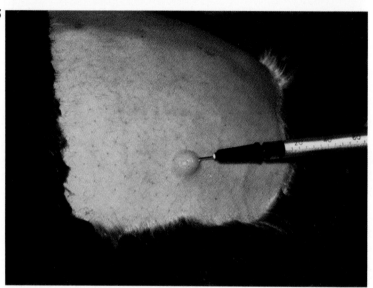

225 Intracutaneous injection. Intracutaneous injection is sometimes given as a route of immunization. Remove all hair in the area by using a depilatory preparation, allowing 24 hours for the skin to recover.

226 Immunization by scarification. The most specific animal antisera are made by natural infection instead of injection. When the skin is scarified and rubbed with a virus suspension an infection will occur followed by antibody production. Scarification is done by special instrument, by scratching with a needle or by rubbing with sandpaper.

227 Scarification and infection with herpesvirus. This rabbit skin was infected with herpesvirus type I after scarification of the skin. In a few days lesions appear, antibodies will be formed in the next weeks. Sometimes a rabbit is lost by paralysis and death.

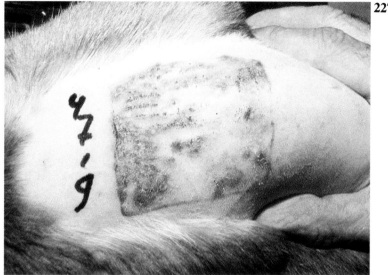

228 Herpes keratitis in rabbits (1). Rabbits' eyes are very susceptible to herpes infections and are used to test antiviral substances. The eyes are infected by giving cocaine anaesthesia, which inhibits the blinking reflex, and applying a virus suspension to the cornea. When lesions develop, in about 3 days, a 20% solution of fluorescein is dripped in the eye. The superfluous fluorescein is washed out and the eyes photographed with electronic flash light.

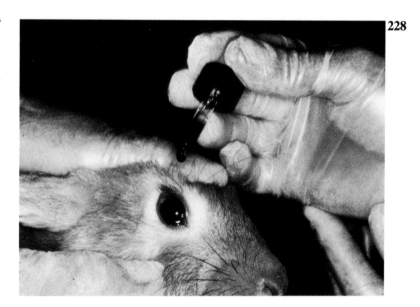

229 Herpes keratitis in the rabbit (2). Herpes lesions in the rabbit cornea are small dotlike epithelial lesions which stain clearly with fluorescein. Photographic recording allows exact registration of the lesions.

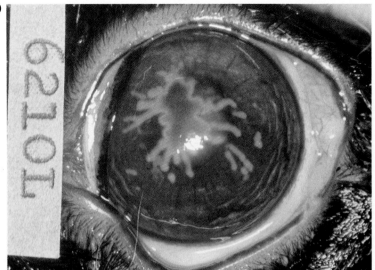

230 Herpes keratitis in the rabbit (3). In a later stage the lesions may get confluent, and keratitis dendritica develops; eventually an ulcer forms.

231 Intranasal infection of a ferret. Ferrets are extremely susceptible to influenza viruses and thus are very suitable for making type-specific antisera. The animal is given narcosis and the virus is dripped into the nose. Keep the infected ferret isolated from the others to prevent spreading of the infection.

232 Orbital puncture in a ferret. Taking blood samples is important to monitor immunization procedures. The puncture is not risky and the amount collected is sufficient for a number of tests. Doing heart puncture endangers the life of the ferret.

233 Monkeys. Monkeys are getting scarce and are to be used with the greatest reservation. In most countries special licences are needed to use monkeys as laboratory animals. The animals are used for erythrocytes (measles haemagglutination), kidney (tissue culture) and for experimental pathology.

COLLECTING BLOOD FOR ERYTHROCYTES

234 Heart puncture in chickens (1). Heart puncture in chickens is easily performed and bears little risk for the animal. In the scheme the exact location of the puncture is indicated, between the two ribs in the angle of the sternum.

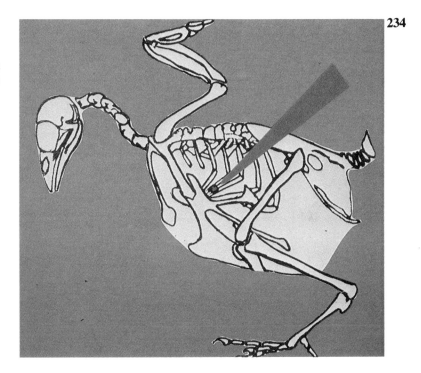

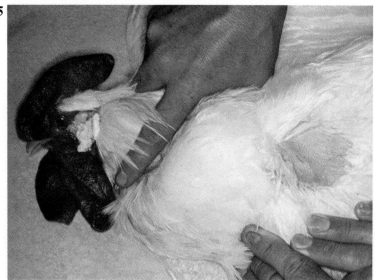

235 Heart puncture in chickens (2). The feathers over the location of the puncture are plucked and the skin is rubbed with alcohol. Most investigators use roosters as their erythrocytes are of a more constant quality.

236 Heart puncture in chickens (3). Some Alsever's solution or sodium citrate is drawn into the syringe before puncture. Thrust in the needle slowly while pulling the plunger. As soon as blood is drawn the needle is pushed another few millimetres to prevent leakage in the pericardial sack.

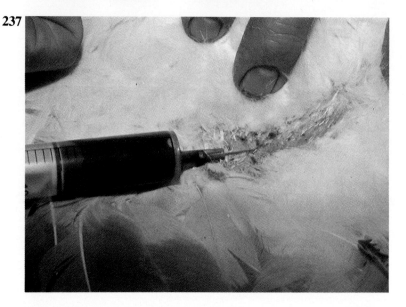

237 Venepuncture in the goose. The goose provides us with large cubital veins which are easily located after plucking away some feathers. Goose erythrocytes are used for rubella and arbovirus tests.

238 Collecting sheep blood (1). The sheep to be bled is laid down on a table and the four legs are tied together as shown here. Anaesthesia or tranquillizing is not necessary.

239 Collecting sheep blood (2). The hair on the skin above the jugular vein is closely cropped and treated with depilatory as only bare skin can be properly disinfected. The depilatory used here is on a barium sulphide base.

240 Collecting sheep blood (3). The blood is collected in the same way as is done in the blood-bank of the hospital. A rubber tube is used to congest the venous system. The blood is collected in Alsevers solution. The washed erythrocytes are used for complement fixation and for the Paul-Bunnell test.

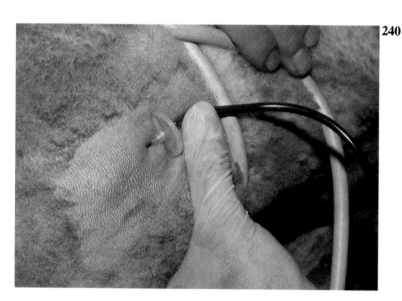

Cell culture techniques

Three kinds of cultured cells are used in virology.

1 *Primary cell cultures*. An organ, or a piece of an organ, is taken and minced into small pieces which are treated with an enzyme solution, trypsin or collagenase, for a relatively short time to free the cells from the tissue. The cells are centrifuged, washed and suspended in a tissue culture medium which supports growth. The cell suspension is filled into bottles and tubes where the cells adhere to the wall and grow out to a monolayer. Sometimes the cells are loosened from the wall by an enzyme solution or E.D.T.A. suspended in medium and made in new cultures. These are secondary cells. Only a few passages can be made from primary cells. The advantage of the use of primary cells is that, if sufficient animal organs are available, any number of cultures can be prepared. The disadvantage is that primary cell cultures can contain viruses that were latently present in the donor animal. Examples are foam viruses and SV40 virus.

2 *Established cell-lines*. These cells can be subcultured forever, they have a malignant character and the number of chromosomes is different from that of the original host. Many of these cell-lines are used. For example HeLa cells, derived from a human cervix cancer; vero cells derived from the kidney of a cercopithecus aethiops monkey; RK13 cells from the rabbit; BHK21 cells from the hamster; and so on. Viruses have different affinities for cells; not every cell-line supports the life of all viruses.

3 *Diploid cell-strains*. Cells can be maintained to be subcultured during 30 to 50 passages and keep their diploid character. The embryonic human fibroblasts are excellent for growing cytomegalovirus and varicella zostervirus which will not grow on any established cell-line. To keep the diploid cell-strain as long as possible, several subcultures are made, the cells of which are stored in liquid nitrogen. Start routine cultures with a high passage, for example the 15th and use the line up to the 30th passage. Then start again with an ampule of passage 15. In this way the cells from the same source can be used for many years giving constant results.

TISSUE CULTURE MEDIA

A large number of tissue culture media have been devised in recent years. Some of them are made for special purposes. Others are made for general use, examples of which are Eagle M.E.M. (minimal essential medium) to be used with addition of serum, and medium 199 which can be used without serum as a maintenance medium. Media can be bought as ready-to-use solutions, as 10 × concentrates or in the form of powders to be dissolved and sterilized by filtration. Sera can be used from many species, for example horse or calf serum are excellent for growing many types of cells. However, they contain antibodies and non-specific virus inhibitors which make them less suited when they are used in cultures in which virus isolations are attempted. Fetal calf serum is free from gammaglobulins and widely used for tissue culture media. Fetal calf serum may contain virus inhibitors, for example for rubella virus, so that serum-free maintenance medium might be indicated. To make up the medium use only water of the purest quality, ultrafiltered or triple distilled, as cells are extremely susceptible for toxic substances.

MONOLAYERS

Cells for virological purposes are mostly used in the form of monolayers. These are formed by the attachment of the cells to the culture vessel wall, stretching of the cells, and multiplication. Normal cells multiply until the cells come into contact with surrounding cells, which inhibit further multiplication. This mechanism is responsible for the formation of the monolayer. Malignant cells lack contact inhibition and the cells stack on each other. Adhesion of the cells to the wall occurs only under strict conditions. The culture vessel wall has to be scrupulously clean and free of toxic substances; it must have the right charge and have a hydrophilic surface. In most cases substances are needed for a good anchorage of the cells. Part of these substances can be secreted by the cells; other substances, like fibronectin, are absorbed into the culture surface from serum.

In other cases the vessel wall must be coated in order to support cell growth. Coating can be done with collagen or polylysine, or by growing endothelial cells on the wall and removing these cells, leaving a layer of so-called extracellular matrix. A specially prepared commercial product is the primairia tissue culture labware (Falcon) in which the protein-like substance is part of the plastic. For normal cell culture procedures in virology the serum used contains a sufficient amount of fibronectin to ensure proper cell attachment. In special cases cell cultures are used which grow in suspensions, the cells of which are not anchorage dependent. To prevent sedimentation of the cells and accompanying metabolic troubles, the medium is kept in continuous motion. There are many systems to do this, from simple rocking bottles to special spinner cultures.

CELL CONTAMINATION OF CELL CULTURES

When cell cultures are contaminated with cells from another cell-line, a mixed monolayer results. If the cells are of a different type this will be observed when the monolayers are inspected with the microscope. However, when the cell strains closely resemble each other morphologically the contamination may go unnoticed.

If a mixed culture of cells is used for subcultures the ratio of the number of cells from each type changes in favour of the one having the best replication rate. In a few subcultures the contaminant cells may have overtaken the original culture. This happens in practice. A few simple rules can prevent these accidents:

Never work with more than one cell strain at a time in a safety cabinet.

Never use pipettes a second time because they are 'only used for sterile work'.

Always use aliquots of medium sufficient for one task, discard leftovers.

Mistrust any change in character of a cell strain.

MICROBIOLOGICAL CONTAMINATION

Cell cultures are ideal surroundings for micro-organisms and, as absolute asepsis is impossible, antibiotics are added to the medium to reduce microbiological contamination. Micro-organisms spoiling cell cultures are bacteria, fungi, viruses and parasites.

Bacterial infection can be prevented by adding a broad acting combination of antibiotics to the medium. Addition of 20 to 40 mg gentamycin and 50 mg vancomycin per litre inhibits practically all bacterial and mycoplasmal growth. Other combinations are possible but only after expert information as some combinations of antibiotics are antagonistic.

Fungal growth is inhibited either by amphotericin B or nystatin. Contamination with viruses comes from the donor animal in primary cell cultures or from biological products used for preparation of cell cultures. Trypsin can contain porcine viruses, and sera, even fetal calf serum, can contain viruses which grow in many kinds of cells. Adding antisera to suppress viral contaminants can reduce replication but will not prevent steady-state infection. Parasites come from the donor tissue. They are seldom a nuisance and no additives are used to suppress their growth.

CELL CULTURE SYSTEMS

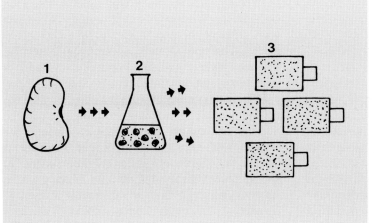

241 Primary cells.

1 Organ from which the cells are derived
2 Pieces of the organ are trypsinized
3 The cells are grown on culture vessel walls

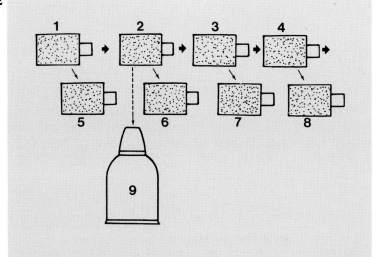

242 Continuous cell-lines.

1 to 4 Continuous propagation of the cells by bi-weekly splitting
5,6,7,8 When the cell-cultures are split, a portion is used to produce cells for daily use
9 At regular intervals cells are stored in liquid nitrogen as a safety measure and as stock for future use

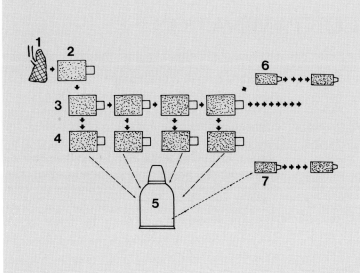

243 Diploid cell-lines.

1 Organ from which the cells are derived
2 Primary culture
3 Subsequent passages
4 From each passage most of the cells are stored in liquid nitrogen
5 Liquid nitrogen storing canister
6 From higher passages side lines are available for use in the laboratory
7 From the nitrogen storage any subculture can be started

PRIMARY KIDNEY CELL CULTURES

244 Primary kidney cell cultures (1). The donor animal is killed by an overdose of anaesthetics. After shaving, the skin is disinfected with iodine. The skin is incised and peeled carefully away.

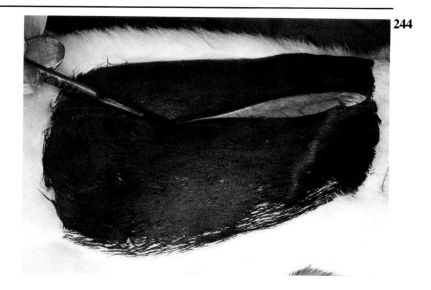

245 Primary kidney cell cultures (2). The muscles of the back are cut through above the region of the kidney. The kidney is taken out aseptically and taken to the laboratory for further handling, which must be done within one hour.

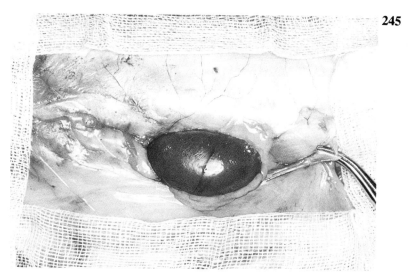

246 Primary kidney cell cultures (3). The outside of the kidney is freed of all adhering tissue. The renal capsule is cut and peeled off. The kidneys are sliced in two halves, using a knife (*not* scissors) to prevent tissue damage.

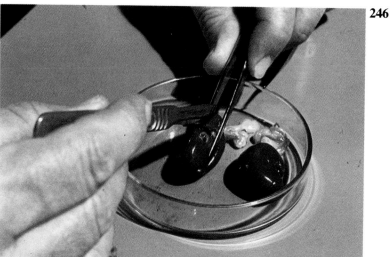

247 Primary kidney cell cultures (4). The pelvis and calices are removed from the halved kidneys, as only the cortex will be used to prepare the cell culture.

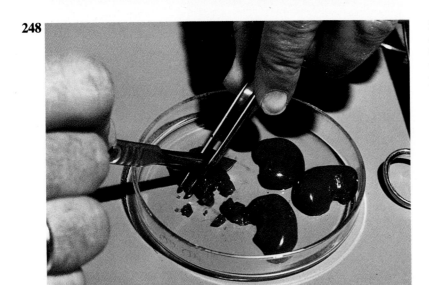

248 Primary cell cultures (5). The cortex is cut with a scalpel into pieces of 3-4 mm. Cutting with a pair of scissors will cause excessive cell loss because of the great pressure exerted on the tissue fragments.

249 Primary kidney cell cultures (6). The kidney fragments are brought from the petri dish into the Erlenmeyer flask and are rinsed several times with saline to remove as much blood as possible. The washing fluid is decanted and discarded.

250 Primary kidney cell cultures (7). After washing the kidney fragments, prewarmed trypsin solution is added. A stirrer magnet is brought into the suspension. Sometimes special trypsinizing flasks are used which have depressions in the wall to enhance the turbulence in the fluid. These flasks are not strictly necessary.

251 Primary cell cultures (8). Trypsinization is performed in a thermostatically controlled water bath. The magnetic stirrers are placed under the water bath. The Erlenmeyer flasks are kept from floating by placing a metal collar around the neck.

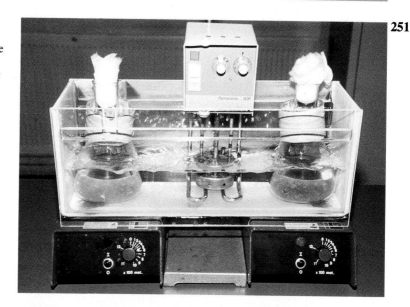

252 Primary kidney cell cultures (9). Every 20 minutes the trypsin solution is decanted and replaced by an equal amount of prewarmed trypsin solution. The decanted fluid, which contains the kidney cells, is collected in an Erlenmeyer flask placed in melting ice. This is done to stop the trypsin activity.

MOUSE EMBRYO CELLS

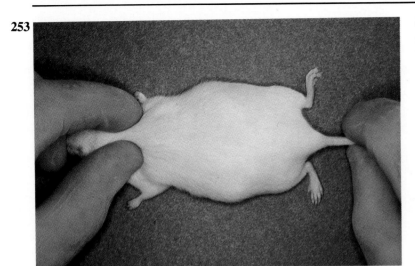

253 Mouse embryo cells (1). A pregnant mouse at 16 to 18 days of gestation is killed by cervical dislocation, as follows. Support the head by placing thumb and index finger behind the occiput. Dislocate the cord by pulling the tail at the root, to prevent breaking off.

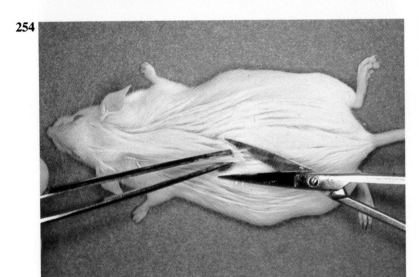

254 Mouse embryo cells (2). To prevent hair and scales flying around the skin is wetted with alcohol 70%. A cut is made in the skin of the back.

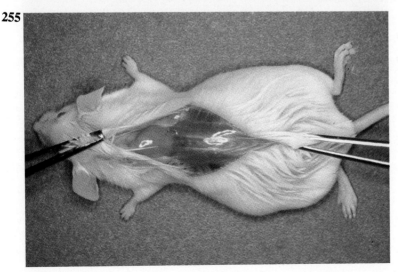

255 Mouse embryo cells (3). Using two surgical pincers the skin is torn away from the back and ventral side in one sweep. In this way the underlying tissue remains sterile.

256 Mouse embryo cells (4). The abdominal wall is opened. The uterus carefully lifted out and cut off just above the pelvis, taking care not to sever the intestines.

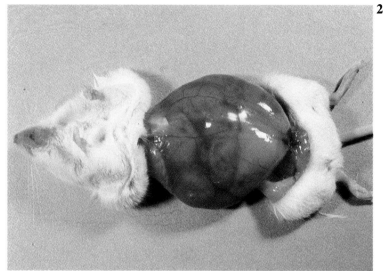

257 Mouse embryo cells (5). The womb is placed in a sterile petri dish. The embryos are taken out by cutting the wall. Discard the empty womb and the placentas.

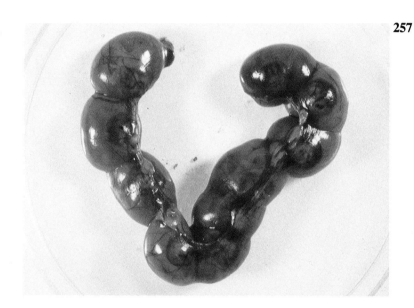

258 Mouse embryo cells (6). Collect the embryos in a sterile petri dish and wash with saline.

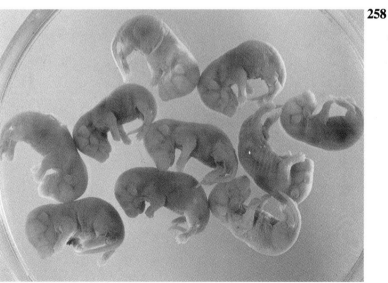

259 Mouse embryo cells (7). Decapitate the embryos and remove the liver and the intestines. Omitting this step will lead to much debris in the cell culture. The cleaned carcasses are washed in saline to remove tissue and blood.

260 Mouse embryo cells (8). A thick-walled tube and scissors are used to mince the mouse embryo carcases for trypsinization, which is carried out as for kidney cells.

CHICK EMBRYO FIBROBLASTS

261 Chick embryo fibroblasts (1). Eight to ten-day incubated eggs are chosen for the preparation of chick embryo fibroblasts. The eggs are candled and the air space marked. After disinfection with alcohol 70% the shell over the air space is removed.

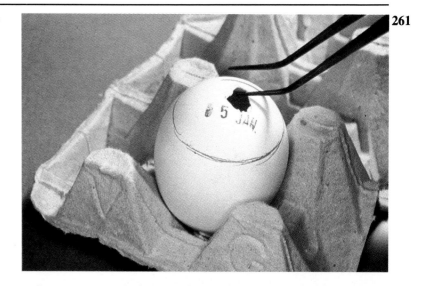

262 Chick embryo fibroblasts (2). The contents of the egg are poured into a petri dish and the embryo is taken out of the amnion sac.

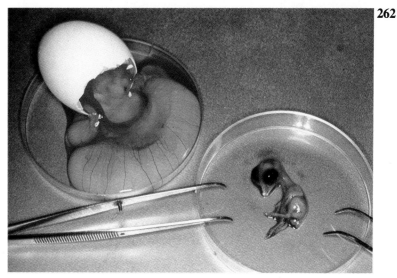

263 Chick embryo fibroblasts (3). The chick embryos are decapitated and the liver and intestines are removed.

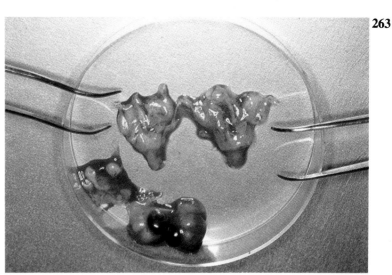

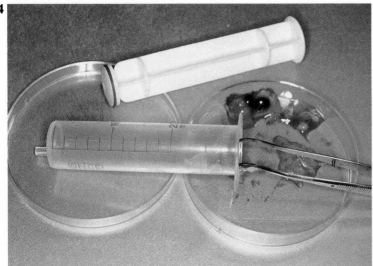

264 Chick embryo fibroblasts (4). Remove the plunger from a 20 ml syringe. Put the carcases into the syringe reservoir, and insert the plunger by pointing the syringe upwards.

265 Chick embryo fibroblasts (5). The carcases are ground by pressing them through the syringe opening. The tissue pulp is washed gently before trypsinization. As the tissue is easily disintegrated, only a few cycles of trypsinization are required.

PERFUSION TRYPSINIZATION

266 Perfusion trypsinization (1). The donor animal (in this case a macacus monkey) is given a deep anaesthesia. The skin is shaved and disinfected. Surgical conditions are used when performing all manipulations.

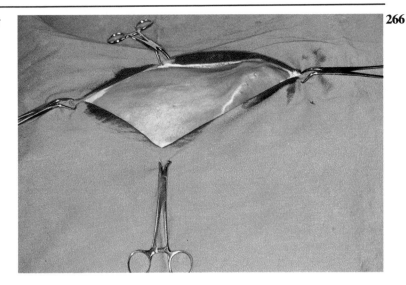

267 Perfusion trypsinization (2). The kidneys are removed under aseptic conditions. Care is taken that the renal artery remains fastened to the kidney as this artery must be used for the perfusion.

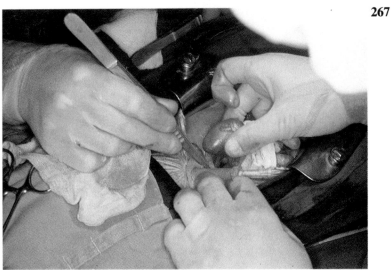

268 Perfusion typsinization (3). For perfusion the kidney is placed in a sterile receptacle and a blunt needle is inserted in the renal artery. The needle is connected to a bottle of trypsine solution which is kept at 37°C. The flow rate is adjusted by raising the bottle to maintain the desired pressure.

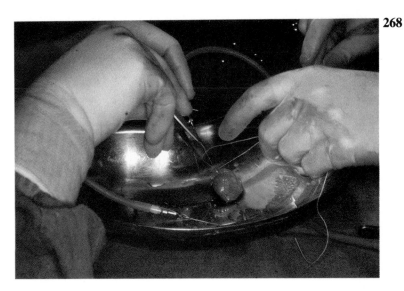

269 Perfusion trypsinization (4). The rate of the trypsinization can be seen at the surface of the kidney as the tissue discolours to yellow with the progression of the process.

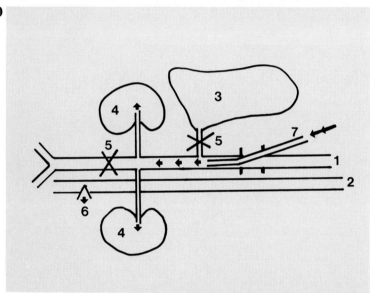

270 Perfusion trypsinization (5). Perfusion trypsinization can be done with the organ in situ and this can have advantages over the surgical method. The aorta is ligated just before the bifurcation. The artery to the liver and the mesenterium is ligated. A cannula is inserted in the aorta just above the diaphragm. The vena cava is severed to allow escape of the perfusion fluid.

  1 Thoracic aorta
  2 Vena cava
  3 Liver
  4 Kidneys
  5 Ligatures
  6 Cut in vena cava
  7 Cannula for perfusion

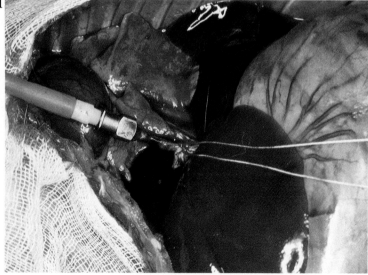

271 Perfusion trypsinization (6). Perfusion in situ at the stage of insertion of the needle into the descending aorta. The animal is under deep anaesthesia but dies as soon as the thorax is opened.

272 Perfusion trypsinization (7). After perfusion the appearance of the kidney is yellowish-white. The capsule is cleaned of adhering fat and connective tissue.

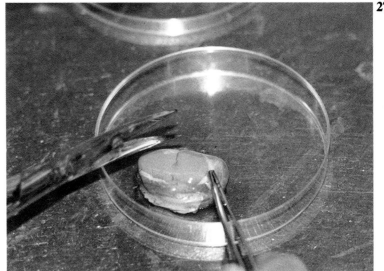

273 Perfusion trypsinization (8). The capsule is cut open and the contents are gently scraped out with a knife. The paste-like mass which still contains a high percentage of aggregated cells is mixed with warmed trypsin and stirred with a magnetic stirrer at 37°C. If the perfusion is less successful an extra cycle of trypsinization enhances the cell harvest.

274 Perfusion trypsinization (9). The trypsinized tissue is collected in an Erlenmeyer flask and cooled down to 4°C. A sterile funnel covered with a double layer of gauze is used to filter the trypsinized tissue into a centrifuge bottle.

275 Perfusion trypsinization (10). Slowly pour the cell suspension onto the gauze; do not stir the sediment. The remnants on the gauze are discarded. The centrifuge bottles are covered with sterile paper and closed by a rubber band.

276 Perfusion trypsinization (11). The filtrate is centrifuged at 800 rpm for 15 minutes in a cooled centrifuge. The sediment is washed and centrifuged twice. After the second time the cells are resuspended in a smaller volume of medium.

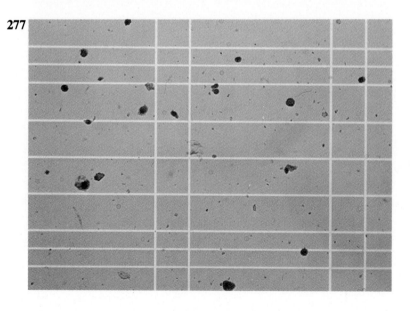

277 Counting cells. Cells have to be counted before the cell suspension is diluted to be seeded in bottles and tubes. Too heavy a suspension causes the wall to be overcrowded, too thin a suspension will never give satisfactory cultures. Count the cells in a haemocytometer using trypan blue to count the viable cells. Dilute to get 100000 to 150000 cells per ml.

278 Freezing cells in liquid nitrogen. Primary cells and cells from continuous lines can be stored in liquid nitrogen. To do this the cells are mixed with a medium containing dimethyl sulphoxide and glycerin. The cooling down of the cells is a critical phase and can be done manually or by a computerized instrument. The ampules are stored under liquid nitrogen in special double-walled vacuum vessels.

279 Thawing cells from liquid nitrogen. Ampules from liquid nitrogen are thawed immediately in tepid water. Always protect your face against glass splinters as minuscule leaks in the ampules allow liquid nitrogen to penetrate, building up high pressure at thawing and causing the ampule to explode.

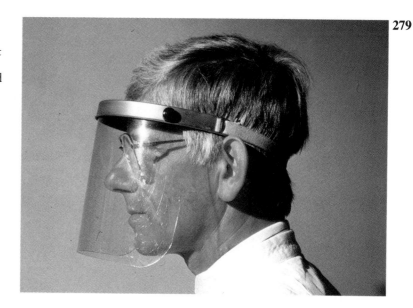

CELL CULTURING METHODS

280 Tissue culture tubes in rack. Tissue culture tubes must be incubated in a slanted position in order to restrict the area of cell growth. Metal racks can be made to order or bought from stock which provide the specific slant for the tubes.

281 The roller drum. The roller drum is a rotating rack for tissue culture tubes. The continuous movement of the tube enhances the replication of certain viruses which results in earlier isolation and higher virus titres.

The racks of the roller drum must be loaded evenly and in balance to prevent a breakdown of the driving mechanism. The roller drum apparatus is placed in a stove of sufficient capacity as the heat produced by the electric motor can influence the functioning of the thermostat.

282 Inspecting cell culture tubes. Cell culture tubes are inspected with low power under a standard microscope. Some provision must be made for holding the tube horizontally. In this case a plexiglass holder is made in which the tube is lying loose to enable shifting and turning of the tube for inspection of the monolayer.

283 A 'Roux' flask. Large surfaces of cells are produced in a pyrex glass 'Roux' flask having a surface of 11 × 21 cm. Disposable plastic flasks of this size are relatively expensive. Glass flasks are readily cleaned and offer a good support for most cell lines. The flask is closed by a non-toxic rubber or silicone stopper. For safety's sake the stopper is taped securely to the neck of the bottle.

284 'Roux' and 'Kimble' glass flasks. A view into an incubator showing tissue cultures grown in large rubber stoppered 'Roux' flasks and small square screw cap flasks ('Kimble' bottles). The small flasks are used to maintain cell lines. The large ones are used for producing enormous numbers of cells necessary for preparing tube cultures.

285 Disposable plastic t.c. flasks. Small (3.5 × 7.5 cm) screw capped tissue culture flasks, made of specially treated polystyrene, are available from a number of manufacturers. These flasks are fit for propagating small stocks of viruses. In most laboratories these flasks are incubated in a CO_2 incubator because of the fact that small gas leaks are not unusual.

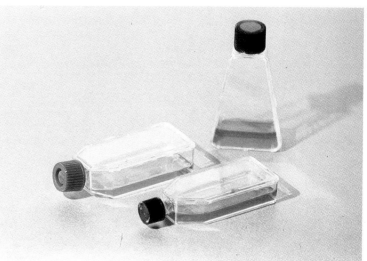

286 Polystyrene t.c. flasks. The colour of the t.c. medium informs us of the condition of the cells. The first bottle shows the light red colour of unused medium of freshly transplanted cells or cells that failed to grow. The second bottle contains medium that turned yellow because of the intense metabolism of the cells. The third bottle had a small leak through which the CO_2 escaped, making the medium alkaline and turning the colour of the medium purplish-red.

287 Prescription bottles. Although prescription bottles are made of soft glass they can be used for cultivating many different kinds of cells. The irregular glass surface can make the microscopic inspection somewhat difficult.

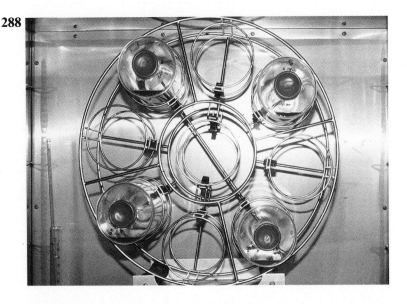

288 Roller bottles (1). Plastic or glass bottles can be used as large surface cell support when they are rolled in a slow-moving roller apparatus. Many different models are made. The one pictured here consists of a wire support in which up to 8 bottles can be supported. Two rubber rollers in the lower part provide the moving power. The apparatus is placed in an incubator at the desired temperature.

289 Roller bottles (2). The cells are grown on the sides of the roller bottle; the amount of medium is relatively small so that high virus titres can be achieved. The level of the medium indicates whether the bottle is perfectly horizontal.

290 Spinner culture. One of the methods of production of large amounts of virus is the spinner culture in which cells are used that will grow in suspension. The medium is agitated by a magnetic spinner driven by the magnetic force of the stirring motor. Screw caps at the top of the bottle allow inoculation of the cells, the taking of samples and the adjusting of the medium.

291 Cytodex beads. Beads of a dextranpolymere (cytodex 1) have a surface that is fit for adhesion of cells in culture. The beads are 50–90 μ in size. One gram of beads provides a surface of 6000 square centimetres, making them fit for large scale production of cells, viruses and antigens.

292 Sterilized cytodex. Cytodex beads are supplied in dry form and are not factory sterilized. Screw cap bottles are used to sterilize 1 g portion of beads in 35 ml of saline. The sterilized beads are washed with medium before use in the culture vessel.

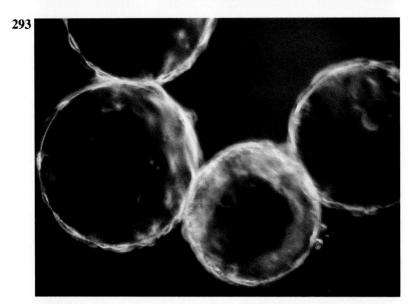

293 Verocells on cytodex beads. Cytodex beads showing adhesion of vero cells. The beads have a density (1.03 g/ml) approximately equal to the medium and must be kept in motion by a magnetic stirrer or a rocker to prevent sedimentation. Cell growth can be controlled by taking a sample and inspecting the beads in phase contrast or darkfield.

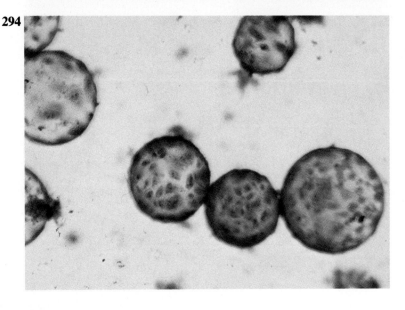

294 Cytodex beads with a coat of vero cells. Stained slides of cytodex beads can be made by taking a sample from the culture, fixing it in 10% formalin and staining the sediment with diluted Giemsa stain. Use a wet mount for microscopic inspection to prevent shrinking of the beads.

295 Leighton tube (glass). The most convenient way to make tissue cultures on cover slips is the Leighton tube. The tube is made of hard glass, contains a square flat bottomed well and is closed with a screw cap. Any cell type which grows on glass can be used.

296 Leighton tube (plastic). As glass Leighton tubes are expensive and difficult to clean many workers prefer the plastic version which consists of a plastic t.c. tube with a flattened side to accommodate the glass 'cover slip'.

297 Leighton cover slip culture. The cover slip is always placed in the Leighton tube in such a way that the cut off corner is in the right upper corner. In this way it is always certain that the tissue culture is on top when the cut off corner is in the upper right side. After staining the cover slip is placed face down on the microscope slide, the cut off corner is then on the top left.

THE COLLODIUM METHOD

298 Making stained slides from t.c. by the collodium method (1). From left to right. 1. The t.c. tube to be used. 2. Tubes rinsed with saline and filled with methanol. 3. Staining with haematoxylin. 4. Rinsed with water and dehydrated in alcohol. 5. Tube filled with collodium. 6. Collodium poured out and after slight evaporation of the alcohol-ether filled with water which precipitates the collodium.

299 The collodium method (2). A glass rod is used to loosen the collodium membrane from the tube wall. The membrane is gently pulled out and immersed in tap water.

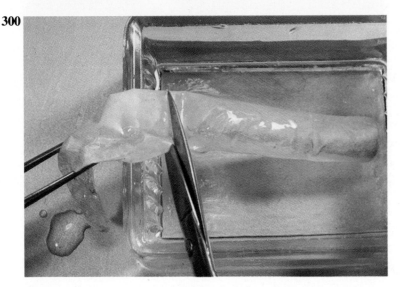

300 The collodium method (3). The cells stained by haematoxylin can be seen in the lower part of the collodium membrane. The part of the membrane containing the cells is cut out and the rest of the collodium is discarded.

301 The collodium method (4). The collodium membrane pieces are trimmed and dehydrated by bathing them in, respectively, alcohol 50%, 70%, 96%. Do not use alcohol 100% as this softens the collodium membrane.

302 The collodium method (5). After dehydration in 96% alcohol the membranes are immersed in 2½% eosin in alcohol 96% for 75 seconds. After this the membranes are differentiated in a few changes of alcohol 96% until the collodium is clear and the cells are still stained, then the membranes are placed immediately in carbol-xylene.

303 The collodium method (6). The collodium membranes are rinsed in xylol and are slid onto a microscope slide. A drop of mountant is applied on the membrane and covered with a cover slip.

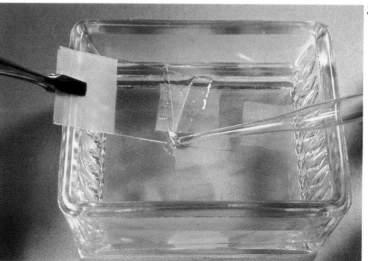

304 The collodium method (7). Close-up of a microscope slide on which is mounted a stained collodium membrane with normal tissue culture cells. The cells form a monolayer with thinning out at the edges.

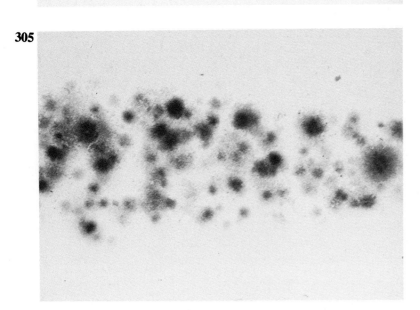

305 The collodium method (8). Collodium membrane containing the outgrowth of malignant cells which grow in islands due to lack of contact inhibition.

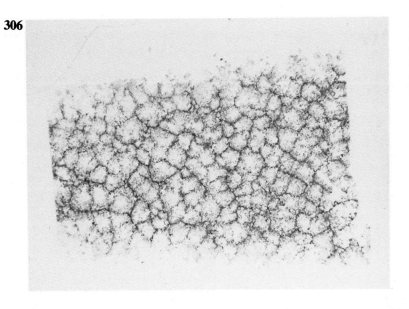

306 The collodium method (9). A monolayer inoculated with vaccinia virus causing focal lesions in which the cells show cytopathic effect. The intense staining comprises the normal cells between the lesions.

307 The collodium method (10). The cell pattern on this slide could easily be mistaken for a cytopathic effect. It is, however, an artifact caused by the fact that the collodium impregnation of the cell layer was imperfect so that not all cells were embedded in the collodium.

CELLS ON SLIDES AND PETRI DISHES

308 Tissue culture chamber/slides. Tissue chamber/slides (this one is made by Lab-Tek) provide the possibility of performing a virus titration, a virus neutralization or a virus typing on a microscope slide. The system is incubated in a carbon dioxide incubator.

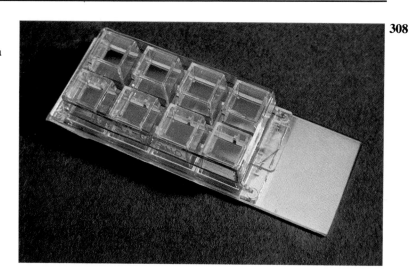

309 Stained tissue culture chamber/slide. To make permanent slides the medium is removed with a pasteur pipette and methanol or 10% formaline are added to fix the cells. The plastic chamber part is removed and the slide stained in the usual way. Here an uninoculated slide is shown.

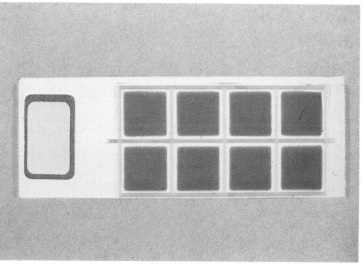

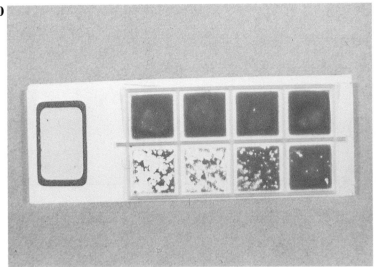

310 Virus titration on tissue culture chamber/slide. A titration of herpes type I virus was made on vero cells in a tissue culture chamber/slide. The lower row was inoculated with dilutions of 10^{-1}, 10^{-2}, 10^{-3} and 10^{-4}, the upper row with 10^{-5}, 10^{-6}, 10^{-7}, 10^{-8}. The cells in the lower row are completely destroyed; the cell layers in the upper row show typical herpes plaques.

311 Petri dishes. Ordinary disposable petri dishes have bottoms which are unsuitable for the adherance of cells, but cover slips can be placed on the bottom as a support for cells. Specially treated petri dishes are available in which cells adhere to the bottom and grow out as a monolayer.

312 Petri dishes, safety measures. Petri dishes with cell cultures, even when not inoculated with virus, must be placed in a large glass or plastic petri dish before placing in the CO_2 incubator. This will protect the incubator against the spilling of medium which will cause bacterial growth.

313 Cell culture in petri dishes. Cover slips from petri dishes can be stained the usual way and be mounted on a slide.

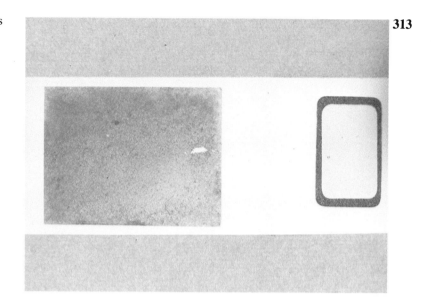

314 Membrane bottomed petri dishes. Petri dishes with bottoms made of a membrane can be used to grow cells on. The bottom permits better gas exchange, but also permits the cutting of the membrane into pieces for staining and fluorescence. This latter is also the case with dishes with a polyester liner (Falcon).

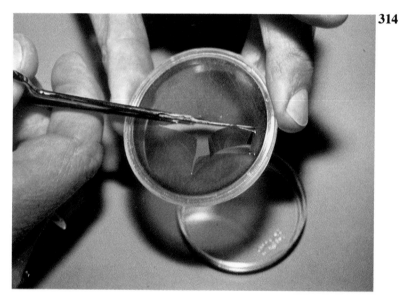

315 Plaque counting. Plaques are virus colonies kept in place by agar containing all the nutrients and neutral red as a vital stain. Where cells are destroyed by virus the monolayer is unstained. These unstained virus-destroyed spots are called plaques. Every plaque contains the descendants of one virus particle so that this technique can be used to count virus, to clone virus (plaque purification) and to assay antibodies (plaque reduction). In the dishes shown a dilution series was inoculated, the right dish remained sterile.

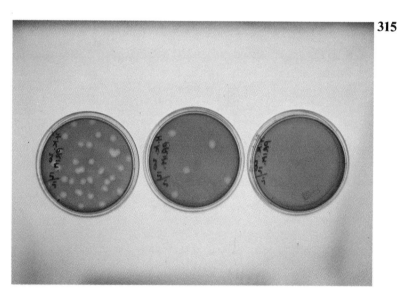

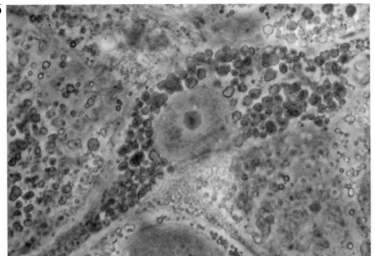

316 Vital staining of cells. The neutral red in the plaques test stains the cells vitally, the cells actively taking up the dye. An example of what happens is shown here.

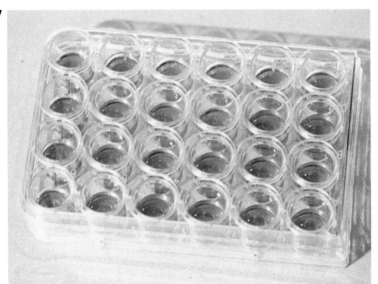

317 Multiwell plastic t.c. plate. Polystyrene tissue culture cluster dish plates offer advantages for many purposes. For example, virus titrations can be done in volumes equal to those of t.c. tubes but with the convenience of reading the results quickly on an inverted microscope. Different kind of cells, media or viruses can be used simultaneously. The volume per dish is sufficient for harvesting. A round cover-slip in a well can be used for making slides for staining.

MICROTITRE PLATE CULTURES

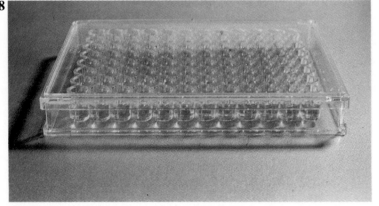

318 Microtitre plates. Microtitre plates with flat bottoms are used for tissue culture. The cell suspension is pipetted into the wells, the plate is shaken, covered with a lid and incubated in a carbon dioxide incubator.

319 Carbon dioxide incubator. The carbon dioxide incubator for tissue culture use is an incubator into which air mixed with CO_2 is pumped. The air is humidified to prevent drying out of the cultures. Simple types use flow meters to regulate the gas mixture; sophisticated incubators are electronically regulated.

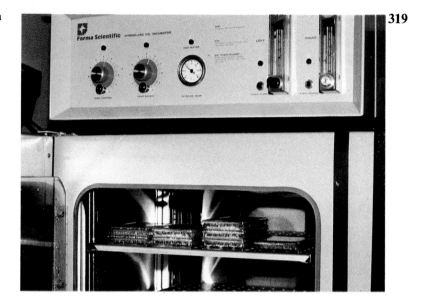

320 Inspecting microtray cultures. Cell cultures in a microtitre tray can be inspected on an inverted microscope. The use of phase-contrast is not necessary. Low power is used to inspect the quality of the monolayer and to detect cytopathic changes.

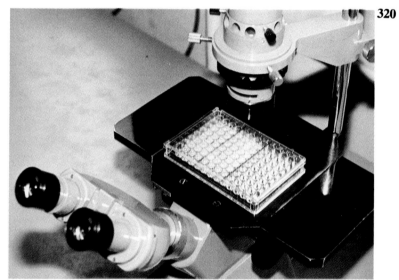

321 Microtitre cell culture – stained. Inspecting the unstained culture well by well is time-consuming. The visualization of fine detail is not necessary. A more efficient way to read the results is to stain the tray with a mixture of formalin 5% and crystal violet 0.13% for 10 minutes. Rinse twice with water and dry inverted on filter paper.

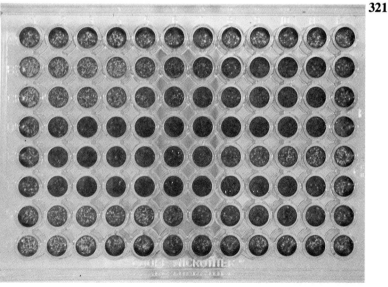

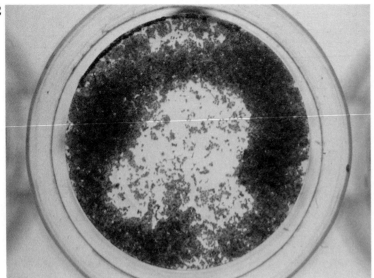

322 Irregular distribution of cells. When the cell suspension is pipetted into the wells, the cells are moved to the rim of the well by the force of the drop. This causes irregular distribution of cells which makes the test hard to read.

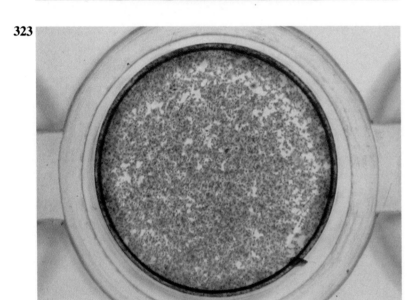

323 Even distribution of cells. After filling the wells of the microtitre plate with cell suspension, the plate is put in a plate shaker for a few minutes to make sure of an even distribution of the cells over the bottom.

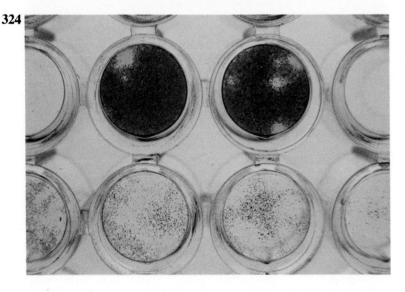

324 Cytopathic effect in microtitre cultures. These wells were inoculated with different dilutions of herpesvirus. Two wells show typical herpes plaques, others have few cells left, some are completely without because of destruction by virus.

325 Titration of viruses. Viruses 1 to 12 were diluted from A to G. The dilution of the virus was made in the microtitre plate and cells were added to all wells. In the wells where no virus was present the cells have formed a monolayer on the bottom. In those wells where virus was present the cells were destroyed and no monolayer was formed.

326 Report of the virus titration. Virus dilutions ranged from 10^{-1} to 10^{-7}; the last row of wells contained the cell controls. The test is easy to read from the stained plate. Virus 1 had a titre of 10^{-5}; virus 3 10^{-6}; virus 4 10^{-2} and so on.

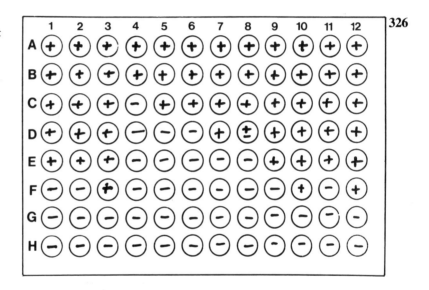

327 Typing virus in microplates. Six Coxsackie B virus isolates were typed as follows. Antisera against Coxsackie B_1 to B_6 were pipetted respectively into row A to F. G contains no serum, H no virus. The strains to be typed were added in duplicate, row 1 and 2 the first strain, row 11 and 12 the sixth strain. After one hour's incubation the cells were added to all wells and the plate was incubated 5 days before staining and reading.

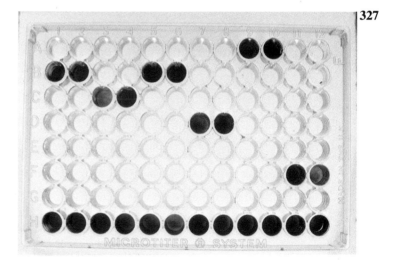

129

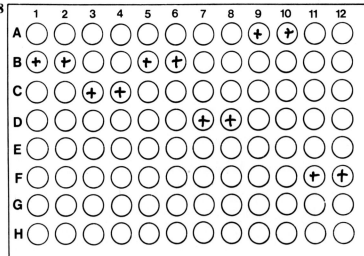

328 Report on the virus typing. In all wells where the virus was neutralized cells were forming a monolayer. So isolate 1 was Coxsackie type B_2; isolate 2 type B_3; isolate 3 type B_2; isolate 4 type B_4; isolate 5 type B_1; isolate 6 type B_6.

SLIDE CULTURES FOR IMMUNOFLUORESCENCE

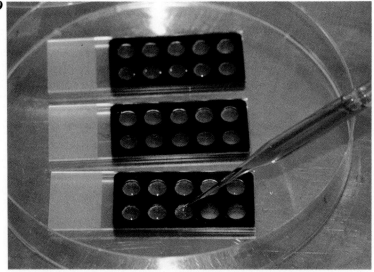

329 Slides for immunofluorescence. Antibodies can be assayed by immunofluorescence on infected cells. The cells can be grown directly on the printed i.f. slides by applying cells in a concentrated drop in each hole, incubating them for one hour in the CO_2 incubator and then adding medium until the cells are immersed. For slow-growing viruses, like c.m.v., the cells can be infected before application to the glass. Rapid-growing viruses are inoculated after 24 hours of cell growth.

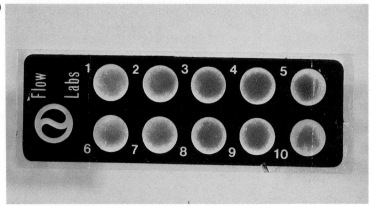

330 Stained immunofluorescence slide. To show the extent of cell growth and to control the c.p.e. the slide was stained with haematoxylin eosin. The cells remain well inside the boundary of the holes; this is important as cells growing on the paint will spread fluid from one well to the other, spoiling the test.

CHLAMYDIAE IN CELL CULTURE

331 Isolation of chlamydiae in cell culture. Chlamydiae, although not belonging to the viruses, are often diagnosed in the virus laboratory. The chlamydiae give rise to eye and genital infections. Smears are put in special transport medium. The chlamydiae are inoculated in HeLa 229 cells growing on round cover slips in flat-bottomed tubes. Infection is assisted by centrifugation.

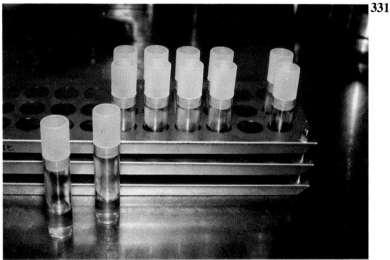

332 Staining chlamydiae. After 3 days of incubation the cover slips are taken out and stained with Giemsa stain or fixed and treated with monoclonal antibodies for immunofluorescence.

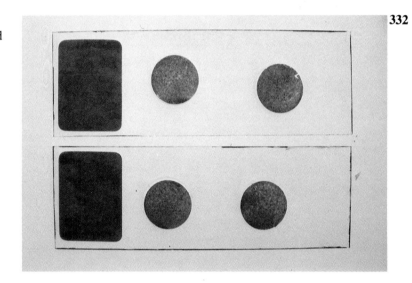

333 Giemsa stained chlamydiae (1). The chlamydial inclusion bodies in the Hela cells are hardly stained by the Giemsa stain. They look like large vacuoles. Confirmation by using higher power and by applying darkfield illumination is necessary.

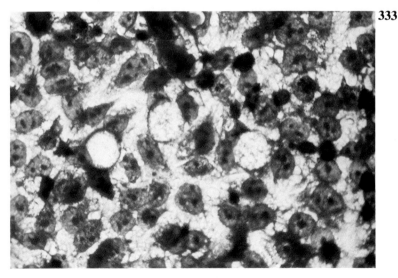

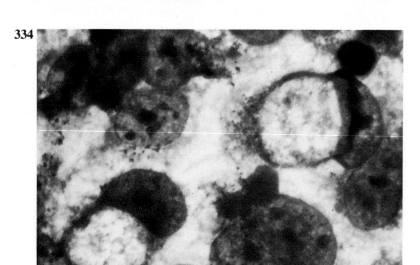

334 Giemsa stained chlamydiae (2). When inspected with an oil immersion lens the 'vacuoles' show all the characteristics of an intra cytoplasmatic inclusion body. The nucleus is pushed aside.

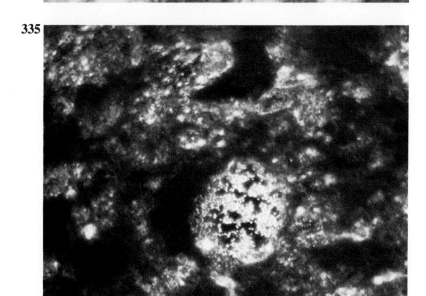

335 Darkfield examination.
When the inclusion bodies stained with Giemsa are examined in darkfield the typical elementary bodies light up.

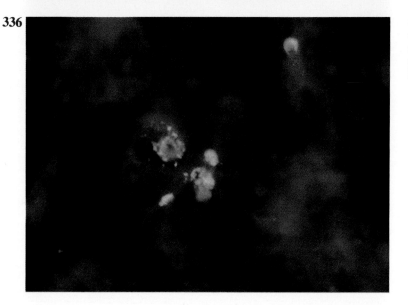

336 Immunofluorescence with monoclonal antibody. Monoclonal antibodies directed against a common chlamydial antigen will specifically detect the inclusion bodies. The background is stained with Evans blue.

Cytopathic effects

Permissive cells in which virus is replicating nearly always show morphological alterations. This is called the cytopathic effect (c.p.e.), being the first sign of virus replication when an infected cell culture is inspected microscopically.

When the cytopathic effect is seen in the unstained culture little information is given on the nature of the virus. Making a stained slide from the monolayer will show details which are helpful for the identification of the virus.

Stained slides can be made either by growing the cells on coverslips in Leighton tubes and petri dishes, or by scraping the cells and making a Sayk sedimentation slide or a cytocentrifuge preparation. Stained slides from culture tubes can routinely be made by the collodium method. Haematoxylin-eosin stain is the one giving satisfactory results in most cases. The cytopathic effect can be diffuse or focal, which may be characteristic of the virus involved. The c.p.e. may be in individual cells or may consist of the formation of giant cells due to cell-fusing properties of the virus.

Inspection of the stained cells can reveal a number of changes which are indicative of the nature of the virus. The changes can be in the cytoplasma, in the nucleus or in both. The alterations seen in the cytoplasma are: a deeper staining because of the enhanced metabolism; retraction of the cytoplasma; rounding of the cell; the forming of an eosinophilic mass, being the site of virus replication or inclusion bodies which may or may not contain virus. These inclusion bodies stain basophilic or acidophilic and can vary in size, shape and number.

Viruses with cell-fusing properties can make neighbouring cells coalesce, resulting in the formation of syncytia. These giant cells can show a number of different characteristics: plasmatic inclusions, vacuoles in the plasma, intranuclear inclusions and combinations of all three.

The nucleus can show all kinds of degeneration, sometimes being pushed aside and deformed by the eosinophilic mass in the cytoplasma. Inclusion bodies of diverse size and density can be observed. In some virus infections the nucleus is enlarged. Although the cytopathic effect is typical for a virus or a virus group, variations may occur due to differences in cells strains and circumstances.

Viruses which show haemagglutination and bud at the cell surface induce the phenomenon of haemadsorption: erythrocytes are added to the cell culture and adsorption to the infected cells – even when c.p.e. is absent – occurs.

Cytopathic effects are changes in the state and appearance of the cells. A number of changes can be induced by circumstances other than virus infection.

A non-specific cytopathic effect can be caused by aberrant circumstances such as a too low or too high incubation temperature which give diffuse alterations with loss of cells. Incorrect composition of the medium as pH, osmolarity, toxic serum or a toxic inoculum cause non-specific changes. With reusable glassware traces of detergent and deposits on the glass wall will result in cell alterations.

One of the most realistic non-specific cytopathic effects is caused by engulfment of particulate matter by the cells. Dead cells, nuclei, materials from an inoculum can all be phagocytized by normal cells and give rise to non-specific inclusion bodies.

ASPECTS OF NORMAL CELL CULTURES

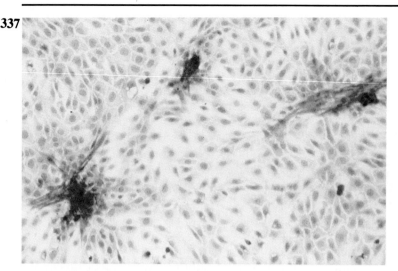

337 Monolayer of mouse kidney cells. Mouse kidney cell culture at the 5th day of the primary culture. The monolayer is completed, the cells show contact inhibition. The dark accumulations are tissue fragments, left over after trypsinization, from out of which fibroblasts grow.

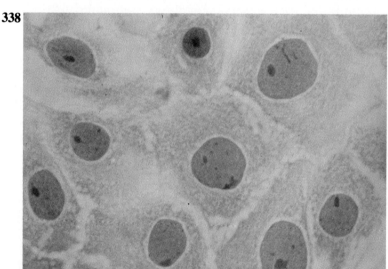

338 Mouse kidney cells. A higher magnification of the cells from the previous picture. the cell borders can be clearly seen due to slight shrinking of the cell layer during fixing and staining. Most of the cells have an epithelial character.

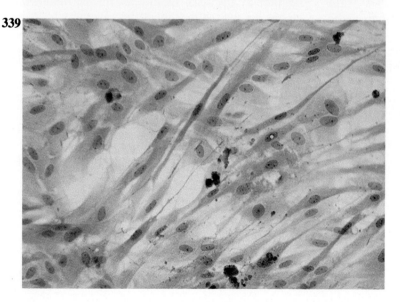

339 Secondary hamster kidney cells. Primary cell cultures can be split several times. The disadvantage is that fibroblasts will overgrow the epithelial cells, as can be seen in these secondary hamster kidney cells.

340 Chick embryo fibroblasts.
Trypsinization of chick embryos results in cultures of chick embryo fibroblasts which have a high rate of multiplication. The picture was made with a phase contrast microscope.

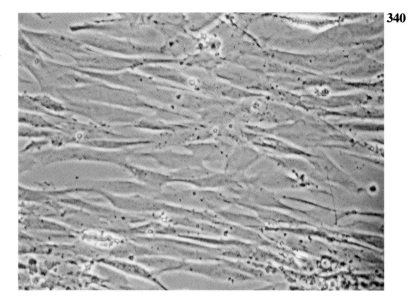

341 Human embryo diploid fibroblasts can be passaged more than 30 times and are able to keep their diploid character. These diploid cell-lines are indispensable to grow varicella-zoster virus and cytomegalovirus. Note the slender appearance of the fibroblasts in this monolayer, which was kept in the roller drums after inoculation with cytomegalovirus. In the central region virus-induced alterations can be observed.

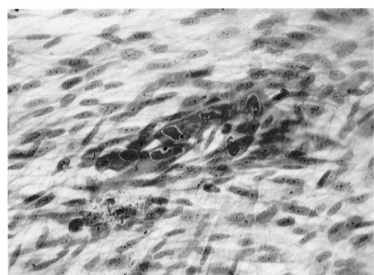

342 Sirc cells. Sirc cells are an established cell line derived from rabbit cornea. The fully grown monolayer here shows the appearance of fibroblastic cells. Rubella virus and herpes virus can be grown on these cells.

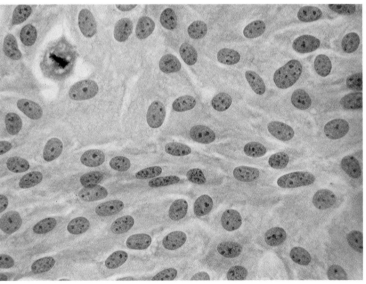

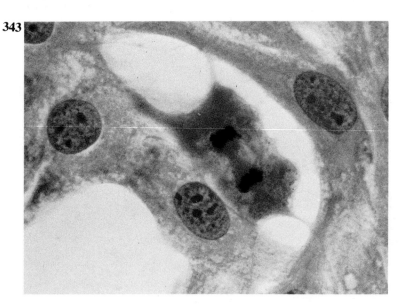

343 Normal cell division. In a young monolayer cell, division takes place all over the sheet. The dividing cell will retract the cytoplasm to round off. Inexperienced observers can mistake cell division for c.p.e. if they look at the unstained, living, monolayer. (Sirc cells).

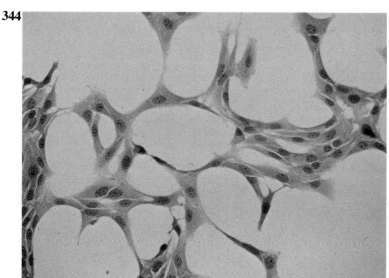

344 Nonconfluent monolayer. When suboptimal number of cells are seeded for a subculture, a confluent monolayer will not be formed, even at prolonged incubation. Here Sirc cells are shown which were seeded too sparsely. Cells reach out for contact with neighbouring cells and stop dividing.

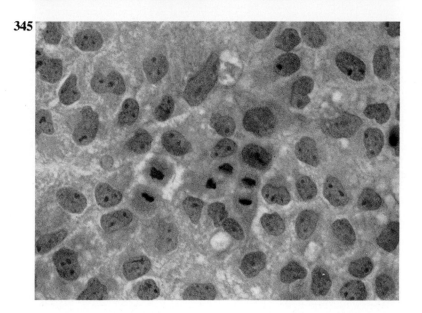

345 BS-C-I cells, young. BS-C-I cells, an established cell-line derived from African Green monkey kidney, have an epithelial character. A two-day-old monolayer is shown here with a lot of normal mitosis.

346 BS-C-I cells, old. Cells have to be split regularly, in most cases twice a week, to keep the cells in optimal condition. The BS-C-I cells shown here were kept for one week and show abnormal mitosis, cell crowding and formation of abnormally large cells.

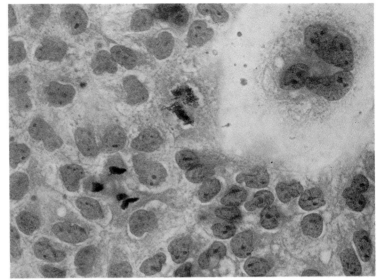

347 Abnormal nuclei. Keeping cells too long in the same culture without splitting will influence mitosis and lead to abnormal nuclei, like the ones shown here in 10-day-old vero cells.

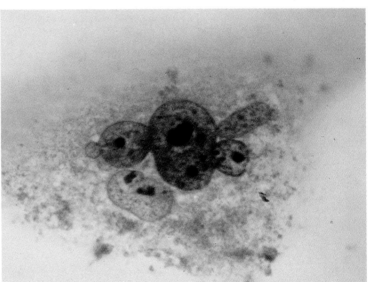

BACTERIAL AND CELL CONTAMINATION

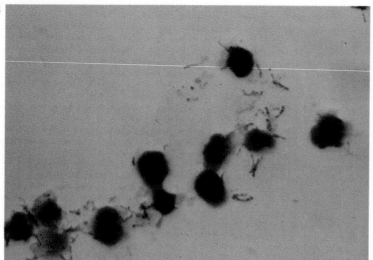

348 Bacterial contamination. Bacterial contamination of cell cultures can be spotted on the more or less turbid medium. Cells are rounding off by the action of bacterial enzymes and metabolites in the medium. The bacteria can be seen in slides from the culture.

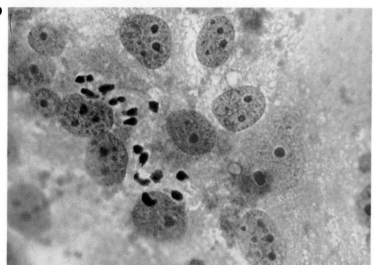

349 T.C. infected with candida. This culture of vero cells was suspected of being infected with candida, although the medium contained nystatin and there was no turbidity. A collidium preparation of the cells clearly shows the yeast infection.

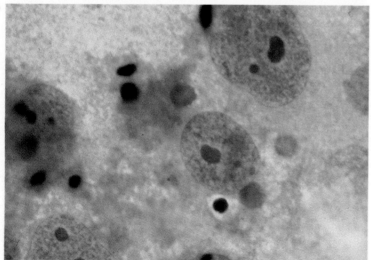

350 T.C. infected with candida. These vero cells show how a cell culture copes with a yeast infection. One yeast cell is phagocytosed, three others are seen in the cell plasma in stages of digestion.

351 Mycoplasma colonies. Mycoplasma is the most silent infection in cell cultures, giving few changes but influencing the quality of the cells. Established cell-lines should be regularly controlled by culture or fluorescence. Mycoplasma grows on special media with very small 'fried egg' colonies.

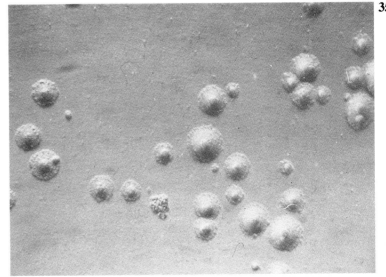

352 Cell contamination (1). One of the most feared contaminations in established cell-lines is cell contamination. The islands in the cell sheet shown here are strange cells, not mutants of the original cell-line. Never handle more than one cell-line at a time. Use aliquots of medium, and discard them after using once.

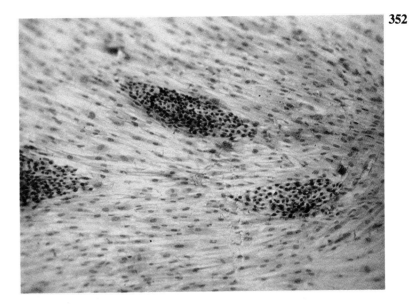

353 Cell contamination (2). This is what happens in cases of cell contamination. The result is not a mixed culture, but one of the cell-lines will overgrow the other and after splitting 2 or 3 times the intruder cells have overtaken the culture.

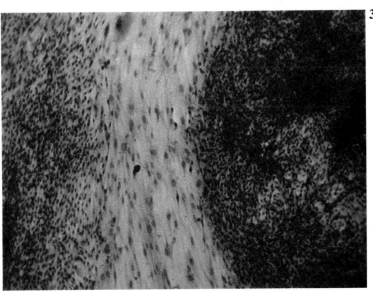

NON SPECIFIC CYTOPATHIC EFFECT

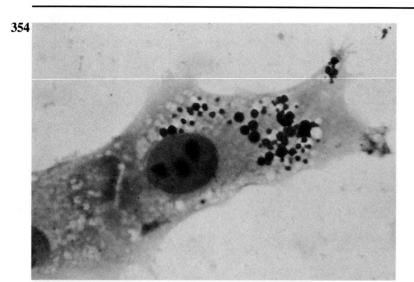

354 False inclusion bodies (1). Cells in culture are able to phagocytize particulate matter. This can lead to false conclusions because of the resemblance to inclusion bodies. This picture shows a Sirc cell which took up yolk globules from the yolk sac material inoculated into the culture. (Giemsa stain.)

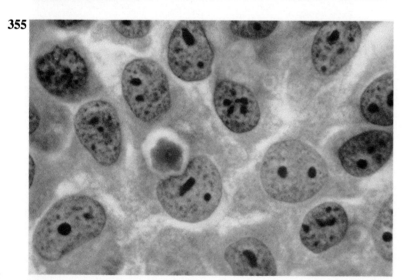

355 False inclusion bodies (2). In every cell culture a number of cells can be observed which contain in their cytoplasma eosinophilic inclusion bodies in a kind of vacuole. This type of inclusion body is caused by phagocytosis of a dead cell or nucleus. The older the culture the more inclusions can be found. (HE stain.)

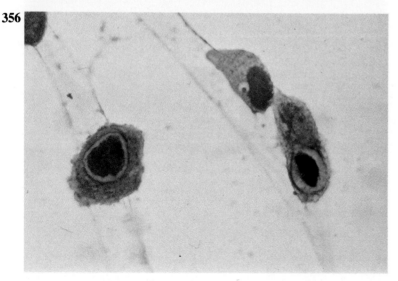

356 False inclusion bodies (3). The intranuclear inclusion bodies, resembling those caused by adenoviruses, shown here are artifacts caused by hypertonic medium. The medium used here was prepared from a concentrated stock solution and a mistake was made during the preparation. The result being a hypertonic medium which caused shrinking of the nuclear content and retraction of the cytoplasm. (HeLa cells, HE stain.)

357 Non-specific cytopathic effect (1).
HeLa cells from an incubator in which the heating failed on Friday evening. Monday morning the monolayer showed this non-virus-induced c.p.e. The temperature which dropped to 18°C was insufficient to allow the anabolic cell processes to take place.

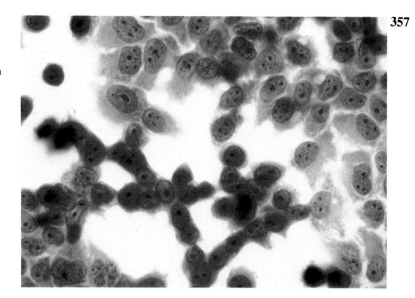

358 Non-specific cytopathic effect (2). A faecal extract was inoculated into tubes of monkey kidney cells. In 24 hours a diffuse cytopathic effect developed involving all cells. This type of cytopathic effect is due to bacterial toxins and further passage into other cell cultures is not possible. Most faecel extracts are non-toxic.

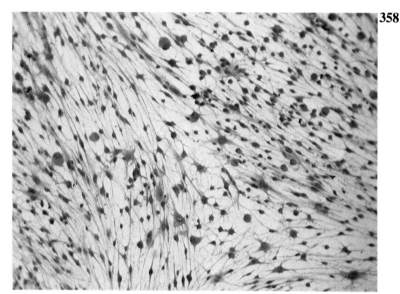

359 Non-specific cytopathic effect (3). The cause of non-specific cytopathic effects can be hard to trace. In this case the changes were due to insufficiently cleaned glassware.

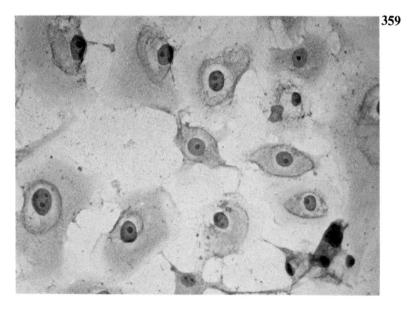

VIRAL CYTOPATHIC EFFECT

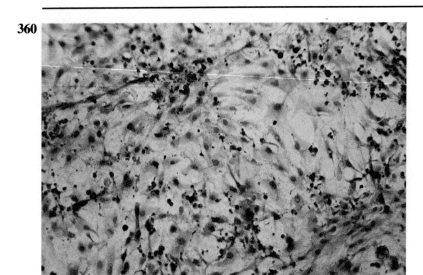

360 Diffuse type of cytopathic effect (1). Viruses which replicate readily and which are released from the cell spread quickly to other cells. The result is a diffuse cytopathic effect all over the cell layer. In the early stages degenerated cells and normal cells occur side by side. (Semliki Forest virus on primary mouse embryo cells.)

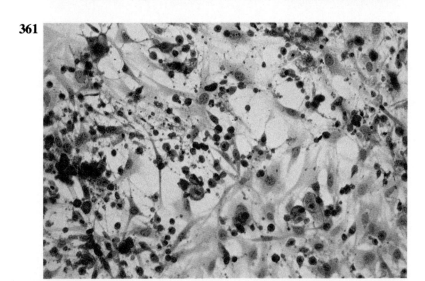

361 Diffuse type of cytopathic effect (2). A nearly complete diffuse cytopathic effect all over the monolayer. In contrast to toxic c.p.e. normal cells still can be found. (Semliki Forest virus on primary mouse embryo cells.)

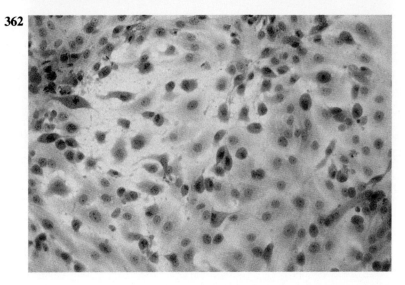

362 Cytopathic effect of Coxsackie B virus. Primary monkey kidney cells 24 hours after inoculation with Coxsackie B virus. The cell sheet shows a diffuse cytopathic effect typical for enteroviruses.

363 The c.p.e. of Coxsackie B6. Coxsackie B6 in primary monkey kidney cells. The c.p.e. is of the diffuse type. The younger stages show different aspects of rounding of the cells, the eosinophilic mass is formed in the cytoplasma and the nucleus is compressed. The later stages show rounding and shrinking of the cells and dark compressed nuclei.

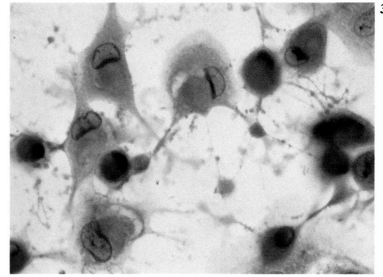

364 Cytopathic effect of poliovirus. Poliovirus grows readily on many kinds of cells. The cytopathic effect is diffuse and consists of rounding of the cells, strong eosinophilic staining of the cytoplasm, and shrinking of the nucleus, which is pushed to the edge of the cell.

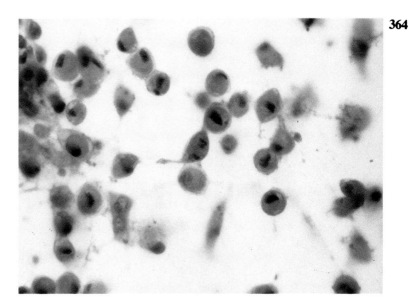

365 Cytopathic effect of ECHO virus. The cytopathic effect of all enteroviruses closely resemble each other. The ECHO viruses give rise to a diffuse c.p.e. Rounding of cells in which the nucleus lies eccentric and plasma with an eosinophilic mass are the main characteristics.

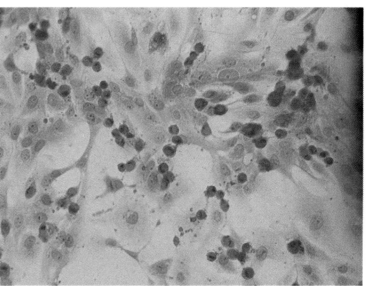

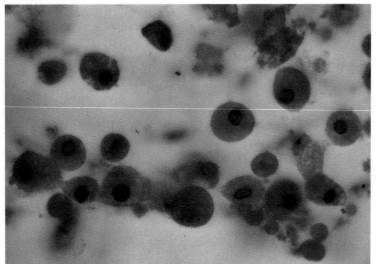

366 Cytopathic effect of Semliki Forest virus. Semliki Forest virus, an arbovirus, gives a diffuse cytopathic effect in primary mouse embryo cells. In the beginning the cells show some vacuolation, later on the cells round off their deeply stained cytoplasm. The nucleus shrinks and is pyknotic.

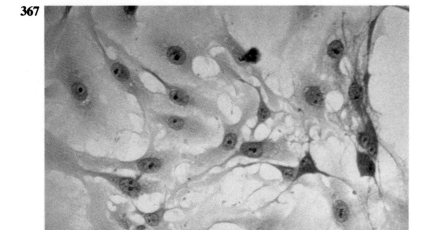

367 Cytopathic effect of West Nile virus (1). The arbovirus West Nile virus replicates well in vero cells. The early c.p.e. is not conspicuous because the cells show only little retraction.

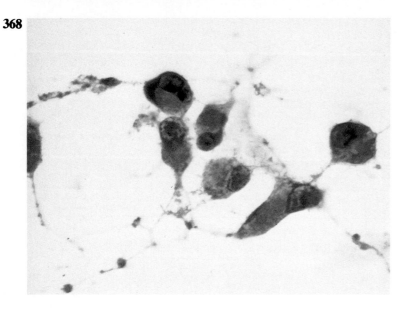

368 Cytopathic effect of West Nile virus (2). The later stage of the c.p.e. of West Nile virus gives rounding of the cells, an intense staining of the cytoplasm and eccentric position of the nucleus.

369 The c.p.e. of encephalomycarditis virus. Early c.p.e. of encephalomycarditis virus (strain Columbia SK) in vero cells. The cytoplasm retracts gradually and rounds off. The nucleus shows typical clumping of chromatin at the nuclear membrane. (Giemsa stain.)

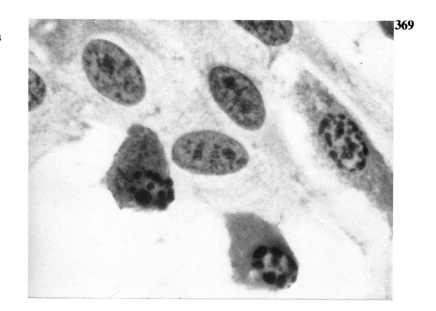

370 The c.p.e. of vesicular stomatitis virus. The cytopathic effect of vesicular stomatitis virus in vero cells consists of cells typically rounded to a pear shape. The nucleus is lying at the rim of the cell; a round eosinophilic mass occupies the centre.

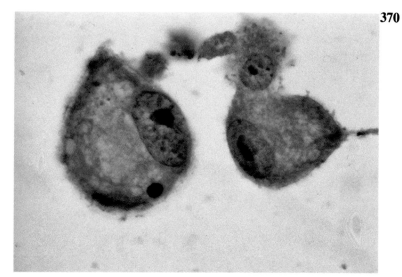

371 The c.p.e. of orfvirus. Orfvirus, causing a pox-like disease in young sheep, can give rise to skin lesions in man. The virus can be grown in cell cultures of fetal sheep kidney. Like all poxviruses, orf replicates in the cytoplasm giving rise to an intensely stained mass.

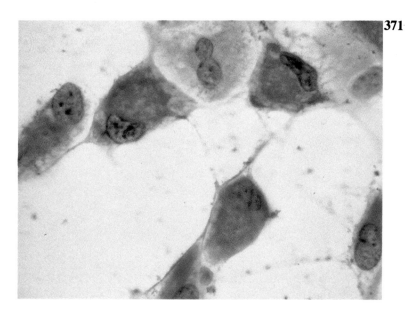

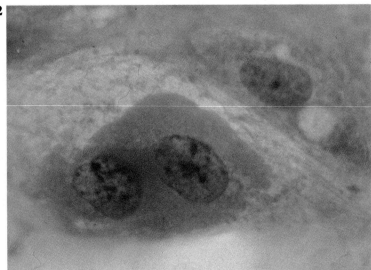

372 Reo viruses. Reo viruses replicate in the cytoplasm where an eosinophilic mass is formed around the nucleus. In the unstained culture cytopathic changes go unnoticed until late in the replication cycle. Reo type III in ferret kidney is shown.

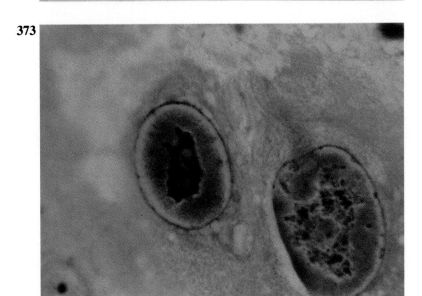

373 Adeno virus type 1. Adeno viruses replicate in the nucleus. The intranuclear inclusion body develops gradually so that in stained preparations all stages can be observed. Adeno 1 in human kidney cells is shown here.

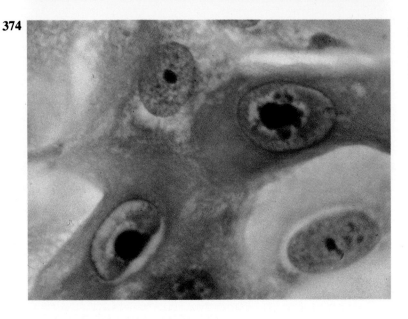

374 Adeno virus type 1. Adeno 1 in human kidney cells showing the late stage of the nuclear changes. A dark mass containing nearly crystalline adeno virus is surrounded by a clear zone. The nucleolus is intact.

375 Adeno virus type 3. The end phase of the cytopathic effect of adeno type 3 in monkey kidney cells. The picture differs somewhat from the c.p.e. shown of adeno type 1. The adeno viruses all have the same type of c.p.e. but with variations which are not specific enough to distinguish the types.

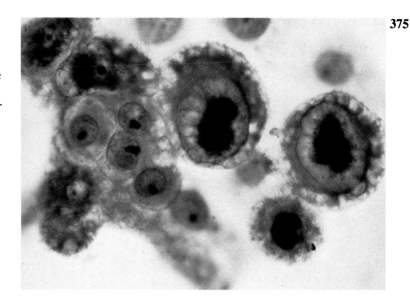

376 Influenza A (1). Influenza A in human kidney. Although influenza virus replicates in the cytoplasma, changes in the structure of the nucleus are obvious. A dark mass in the nucleus is formed early. A haemadsorption test at this stage is more conclusive than the cytopathic effect.

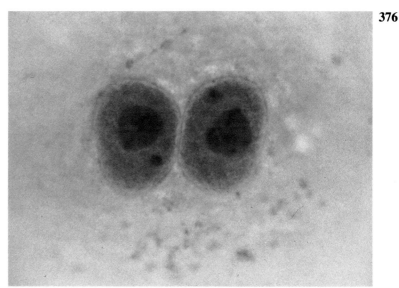

377 Influenza A (2). Influenza A in human kidney, the late stage of the cytopathic effect. The nucleus contains a dark mass surrounded by a light zone. The cytoplasma shows few changes and does not retract. Immunofluorescene will show the virus produced in the plasma.

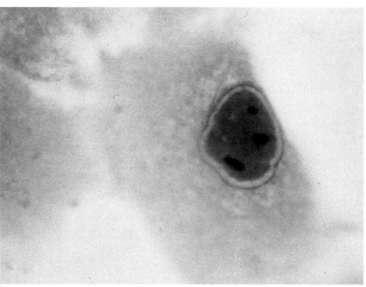

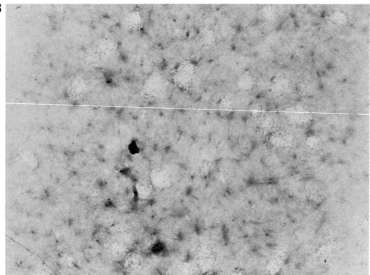

378 Focal cytopathic effect (1). The monolayer of monkey kidney cells was inoculated with herpesvirus type 1. This virus spreads from cell to cell causing infection foci manifesting as clear spots in the monolayer. The number of holes correlates with the number of virus particles inoculated into the cell culture.

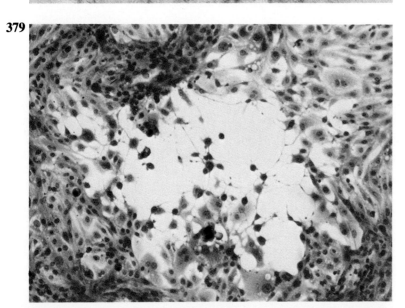

379 Focal cytopathic effect (2). A detail from a monolayer of monkey kidney cells inoculated with herpesvirus type 1. A lesion like this is called a plaque and develops from one cell infected by a virus particle.

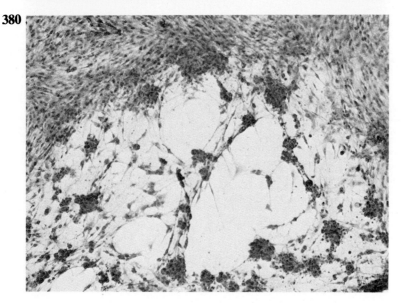

380 Focal cytopathic effect (3). A monolayer of primary monkey kidney cells inoculated with vaccinia virus. In the centre the remaining cells shows maximal c.p.e., while most of the cells came off the glass wall. The rim of the lesion shows diverse stadia of cytopathic effect while the adjoining cells are normal. The cell layer can be covered by soft agar to retain the degenerated cells and the virus produced in them. In this way plaque counts are more exact.

381 The c.p.e. of herpesvirus type 1 in B.S.C. cells. There is always some variation in the type of c.p.e. of a virus, depending on strain variations and on the type of cells used. Here the c.p.e. of the herpesvirus results in rounding of the cells and the formation of a diffuse nuclear lesion (ground glass nucleus) instead of a distinct inclusion body.

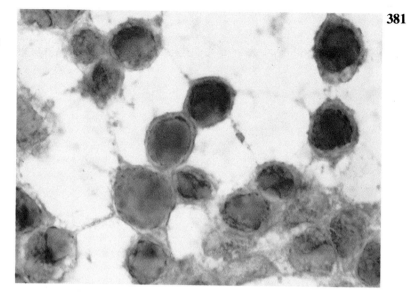

382 HeLa cells infected with H.S.V. type 2. Individual infected cells and secondary giant cells can be observed. Nuclear changes are prominent. The central mass has a typical ground glass aspect. (HeLa cells, HE stain.)

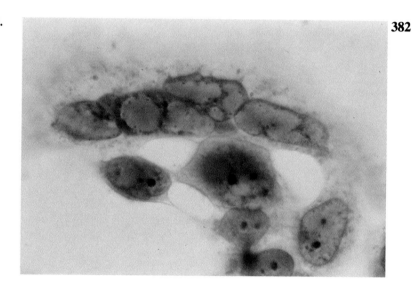

383 Cytopathic effect of varicella virus in a culture of human diploid cells. Irregularly formed intranuclear inclusions develop before the rounding of the cells. Identification is by serological techniques.

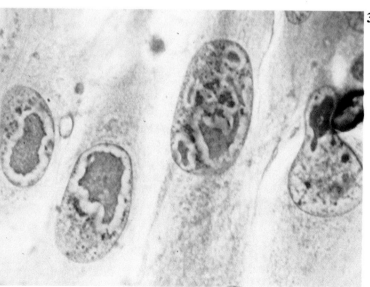

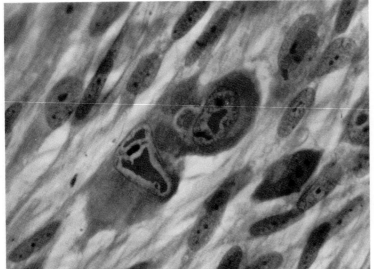

384 Cytomegalovirus cytopathic effect. Human cytomegalovirus replicates only in diploid human cells. The cells become enlarged (hence the name cytomegalo). The cytoplasma stains intensely and can contain an inclusion-like mass. The nucleus is also enlarged and contains one or more irregular dark staining inclusion bodies which contain the virus. The nucleoli remain intact. The cytopathic effect shown here is fully developed.

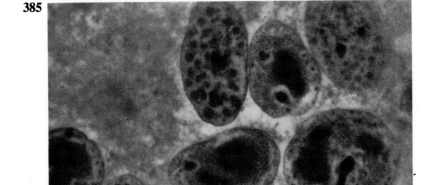

385 Cytomegalovirus – early effects. The nuclei of these human kidney cells show early changes caused by cytomegalovirus. the dark foci in the nucleoplasma are the centres of virus replication.

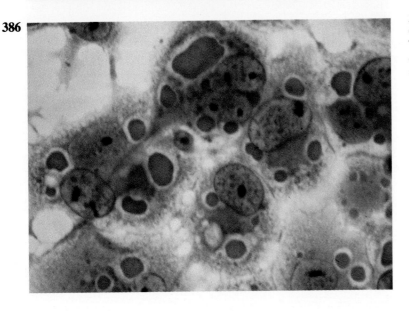

386 Cowpox virus inclusion bodies. Cowpox in primary monkey kidney cells. The cytopathic effect consists of round cytoplasmic inclusion bodies with a halo in single cells and small syncytia.

387 Cowpox inclusion body. This phase contrast photo of an inclusion body of cowpox in a monkey kidney cell shows that the inclusion body is full of small round elementary bodies. These are the virus particles.

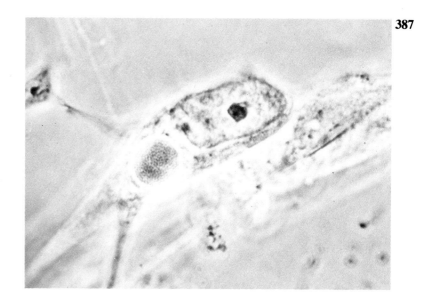

388 Vaccinia virus. Vaccinia virus in monkey kidney cells. The virus replicates in the cytoplasma and inclusion bodies are small or absent. The cytopathic effect of vaccinia virus is focal.

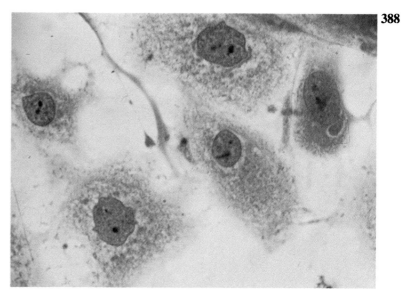

389 Early nuclear changes by SV40 virus. SV40 virus, papova virus, replicates in the nucleus of the cell. The nucleus of this monkey kidney cell has more than 20 foci of virus activity. The cytopathic effect of the human BK virus is closely similar.

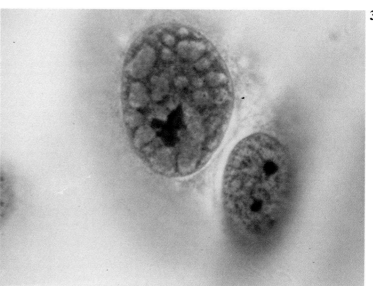

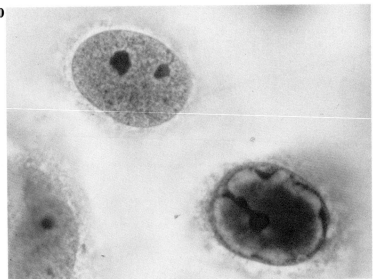

390 Cytopathic effect of SV40 virus. The fully developed cytopathic effect of SV40 virus is restricted to the nucleus in which a dark inclusion body can be seen surrounded by a halo. Some chromatin adheres to the wall of the nucleus. The nucleolus is intact.

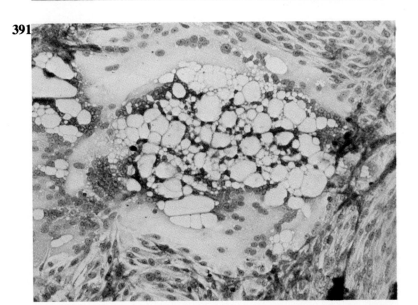

391 The c.p.e. of foamy virus. Spontaneous c.p.e. of a foamy or syncytial virus in a culture of Cynomolgus monkey kidney cells. Foamy virus is one of the simian viruses which contaminate cell cultures of monkey kidneys. The frothy appearance has led to the name spuma viruses for this group of viruses.

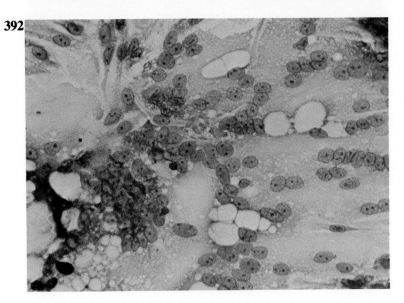

392 Detail of foamy virus c.p.e. In this detail from a foamy virus cytopathic effect it can be clearly seen that it is basically a syncytium, with cytoplasmatic or nuclear inclusions. This is a more constant characteristic of this group than the vacuoles.

393 Syncytial virus from a chick embryo. All animal organs used for making cell cultures can contain indigenous viruses. This syncytial virus was isolated from primary chick embryo fibroblasts.

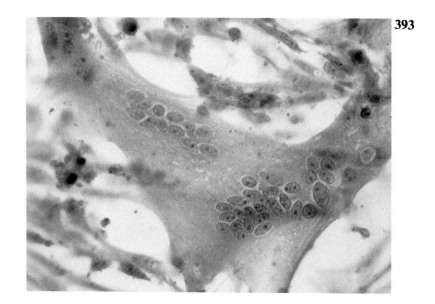

394 The c.p.e. of measles virus. Measles virus in primary monkey kidney cells. The replication of measles virus causes three typical changes in the monolayer: giant cell formation, eosinophilic cytoplasma inclusions, and eosinophilic nuclear inclusions. The cytoplasmic inclusions show clearly in this picture.

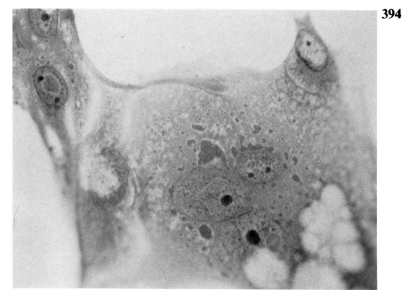

395 The c.p.e. of measles virus. In this view of the cytopathic effect of measles the giant cell character and the intranuclear inclusions are well marked.

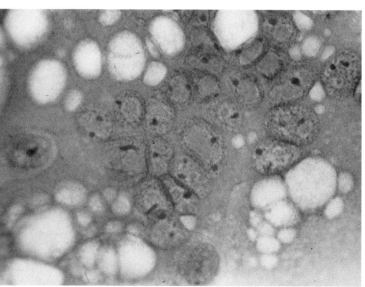

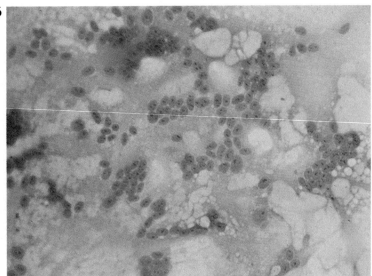

396 S.S.P.E. virus. The virus isolated from a case of subacute sclerosing panencephalitis (S.S.P.E.) causes a cytopathic effect with large giant cells and vacuolation. Antigenically the S.S.P.E. and measles viruses do not differ.

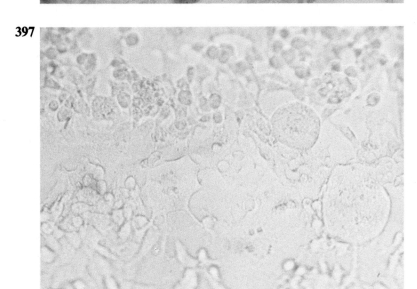

397 Respiratory syncytial virus. The respiratory syncytial (R.S.) virus is named after the syncytia formed in tissue culture. The unstained c.p.e. shown here was formed in HeLa cells.

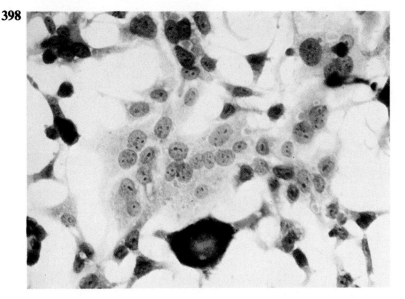

398 R.S. virus. Three forms of the cytopathic effect of R.S. virus in HeLa cells are seen here: cytopathic effect in single cells, a newly formed giant cell and a dark stained giant cell in the end stage of the c.p.e.

399 R.S. virus inclusion bodies.
Respiratory syncytial virus induces the formation of cytoplasmic inclusion bodies in single cells and giant cells. No intranuclear inclusions are formed.

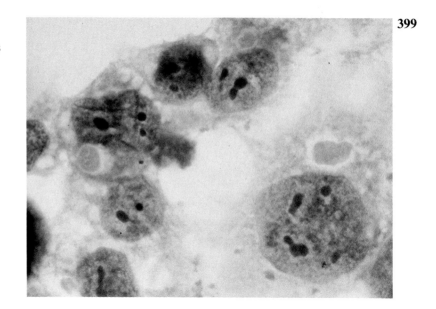

400 Infectious bovine rhino-tracheitis virus. Cytopathic effect of infectious bovine rhino-tracheitis virus in bovine embryo kidney cells. The cytopathic effect consists of small giant cells as well as changes in single cells with clearly visible nuclear inclusions. No cytoplasmic inclusion bodies are formed.

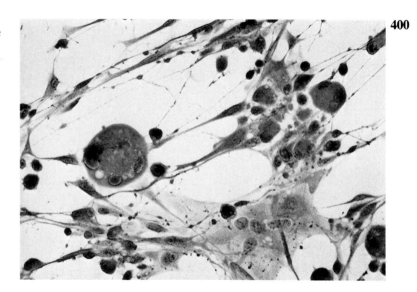

401 Equine herpesvirus. Like all herpesviruses the cytopathic effect of equine herpesvirus in bovine embryo kidney cells is characterized by intranuclear inclusion bodies. Single cell c.p.e. and giant cells both can be observed.

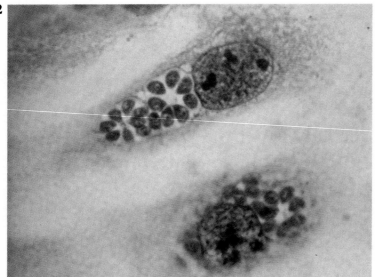

402 Toxoplasma gondii. Toxoplasma gondii grows well in many kinds of cells. The replication of the parasite takes place in vacuoles with a clear wall in which the rosettes of the trophozoids can be seen. The cytopathic effect in this stage is not clearly observed when unstained. Ultimately the cell sheet undergoes total destruction.

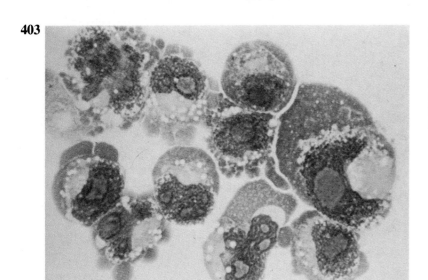

403 Epstein-Barr virus. The Epstein-Barr virus replicates only in human lymphocytes. The cells are transformed and can replicate without limit. The production of E.B. antigens varies from cell strain to cell strain. This slide was made by using a Sayk sedimentation chamber and Giemsa stain.

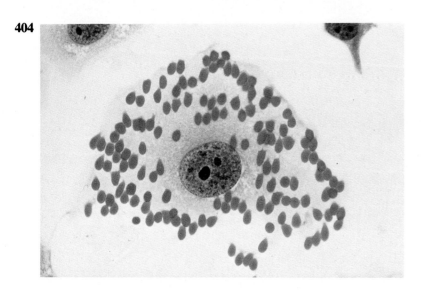

404 Haemadsorption (1). Mumps growing in primary monkey kidney cells. Myxoviruses and paramyxoviruses replicating in tissue culture show the phenomenon of haemadsorption on the surface of the cells which they infect. A suspension of red blood cells, guinea pig or rooster, is added to the culture. Twenty minutes later the cells are rinsed with saline and inspected. Erythrocytes are adsorbed to the infected cells.

405 Haemadsorption (2). Influenza A in primary monkey kidney cells. Orthomyxo and paramyxoviruses can be detected early, even in absence of a clear cut c.p.e., by the haemadsorption technique.

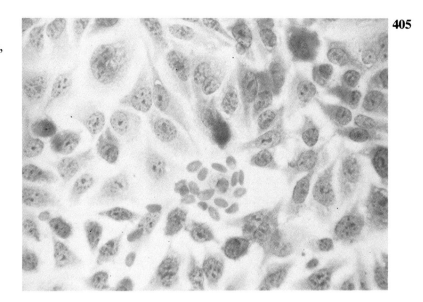

406 Haemadsorption (3). Sendai virus in primary monkey kidney cells. This heavily infected monolayer shows cells with extensive haemadsorption. Identification of a virus can be made by a haemadsorption inhibition reaction, by treating the cells with an antiserum prior to the addition of the red blood cell suspension.

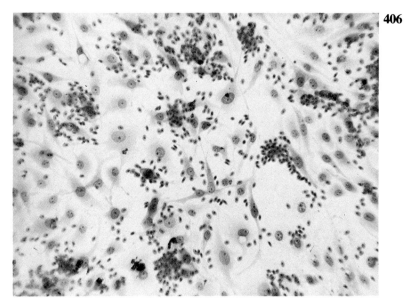

Histopathology of viral infections

In many cases a histopathological diagnosis from a biopsy or an autopsy is necessary to make or to confirm the diagnosis. The specimens for histopathological diagnosis are put into a 10% formalin solution to fix the tissue as soon as possible to prevent autolysis and artifacts. It is important to instruct the clinician in charge of a biopsy, or the pathologist in charge of an autopsy, that further specimens are needed for virus isolation and that these specimens must be sent in virus transport medium. If this is omitted a confirmation of the direct diagnosis is impossible.

Studying the histopathology of virus lesions will give much information concerning the diagnosis and the pathogenesis of viral infection. A number of viral infections are still diagnosed by inspecting stained tissue sections or smears. For example, one of the signs for the diagnosis of mononucleosis is the blood smear in which atypical lympocytes can be observed, although sometimes a severe cytomegalovirus infection can mimic the picture.

Cytomegalovirus infections, which occur in a severe form in very young children and immunocompromised patients, can be studied by examination of sections of the different organs in which the virus manifests itself in the form of typical inclusion bodies.

Molluscum contagiosum is a small, benign, umbilicated papule caused by a poxvirus. The lesions are typical but the diagnosis can be confirmed by excision of the papula and making stained sections. The typical structure of the lesion and the presence of swollen cells with an intracytoplasmic inclusion body are unique features of this disease. Warts are caused by an infection with a papilloma virus. The histology of the lesion comprises proliferation of the epithelium and hyperkeratosis. Herpes simplex, varicella and zoster lesions manifest themselves as vesicles in the skin. The lesion begins as a local infection of the skin in which lysis of the cells starts the process of vesicle formation. Progressive degeneration of cells and transduction of fluid form the typical vesicle the bottom of which comprises cells with nuclear inclusions. A stained smear from this part of the lesion shows whether the cause is an infection with a herpesvirus or not.

The cause of death due to pneumonia in influenza patients can be investigated by autopsy and histopathological studies. There is a marked difference between a pure virus infection and a virus infection complicated by secondary bacterial infection. Studying the morbid anatomy of these cases is of great educational value.

Varicella pneumonia in adults is a common complication of the disease which now, thanks to adequate chemotherapy, is no longer a cause of death. Respiratory syncytial virus and measles virus both cause giant cell pneumonia which can be fatal. In sections of the lung typical giant cells can be found due to the cell-fusing property of these viruses. Lung lesions due to cytomegalovirus are found in perinatally infected children and adults under intense immunosuppression for example kidney transplantation patients. In severe cases, in which a rapid diagnosis is necessary, a lung biopsy is taken to make a range of diagnostic investigations – immunofluorescence, stained sections and virus cultures.

Infections of the brain and spinal cord show appearances typical of the virus involved. For example intranuclear inclusion bodies in neurons and oligodendrocytes in herpesvirus, intracytoplasmic inclusions – Negri-bodies – in infections with rabies virus. In meningitis there is a considerable infiltration of the leptomeninges with mononuclear cells. From there the cells migrate along the blood vessels into the nervous tissue. This is seen in sections of the brain tissue as perivascular infiltrates or perivascular cuffing. In the surrounding tissue scattered inflammatory cells can be seen. Neurons which are infected by virus degenerate and are cleared away by phagocytic cells, microglia and polymorphonuclear neutrophils. The process is called neuronophagia.

A number of viruses cause histopathological changes in the liver. In some cases a needle biopsy is done to make stained slides or to perform immunofluorescence. Changes in cell appearance and inclusion bodies characteristic of the viruses can be found. In fatal disseminated cases of herpesvirus coagulation necrosis with typical nuclear inclusions in oesophagus, liver, adrenals and liver are regularly seen.

407 Normal human liver. Many hundreds of thousands of lobules make up the human liver. The pattern of architecture is always the same for all lobules, although there is considerable variation in size and shape.

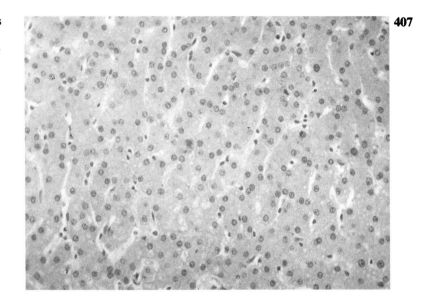

408 Hepatitis (1). Hepatitis gives a diffuse parenchymal infection. The individual hepatocytes are commonly swollen; many of them show ballooning. The degeneration is not localized in any part of the liver lobule, in contrast to yellow fever.

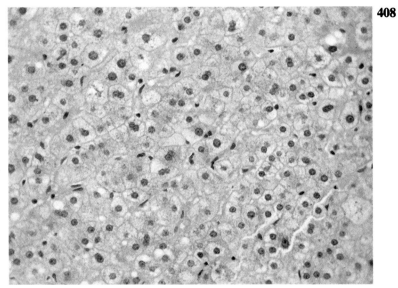

409 Hepatitis (2). The cytoplasm of the swollen cells shows a granular ground glass appearance. The nuclei vary considerably in size and staining, which is pathognomonic.

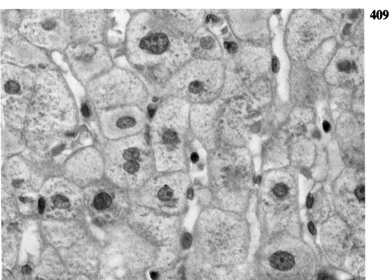

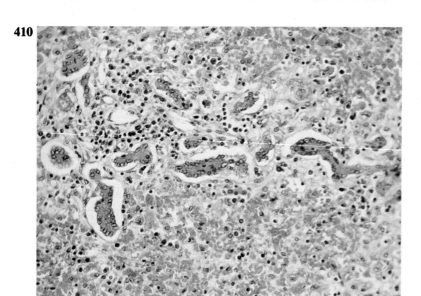

410 Hepatitis (3). A section from a fatal case of hepatitis. There is a diffuse parenchymal destruction through cytolytic necrosis and disappearance of individual hepatocytes. Proliferation of the bile ducts is clearly demonstrated.

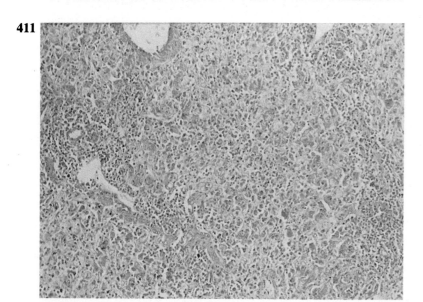

411 Yellow fever (1). Yellow fever, caused by a mosquito-borne arbovirus, is characterized by degenerative lesions of the heart, kidney and liver. The lesions in the liver comprise midzonal necrosis of liver lobules, which means that the lesion is localized between the portal vein and the central vein.

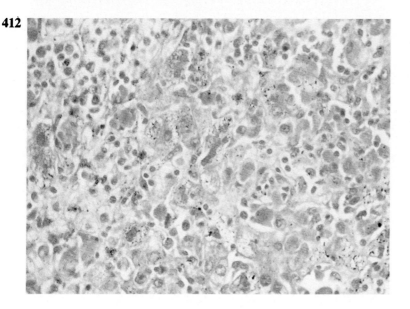

412 Yellow fever (3). The midzonal necrosis, with many degenerated hepatocytes, shows preservation of the basic liver architecture. The mortality of this grave disease is about 10%.

413 Herpes simplex infection of the liver.
Neonatal herpetic infection can run a fatal course, involving many organs. The herpes infection of the liver shows enormous destruction in which nuclei can be found with the characteristic intranuclear inclusions.

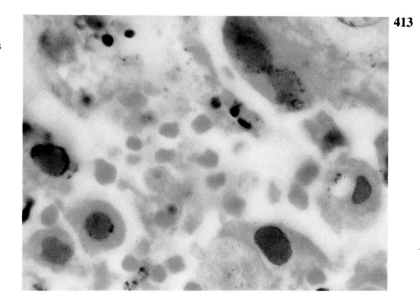

414 Cytomegaloinfection of the liver.
Hepatomegaly, considerable enlargement of the liver, is one of the characteristics of congenital cytomegalovirus infection. Cytomegalovirus causes swelling of the infected cells and characteristic intranuclear inclusions. The typical cells are nicknamed: owl's eye cells.

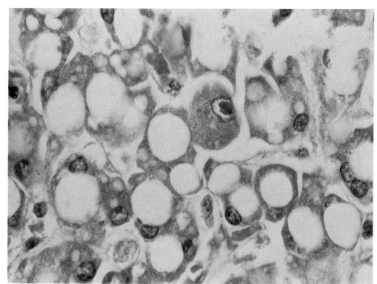

415 Liver lesion in mononucleosis.
Involvement of the liver can be one of the complications of an infection with E.B. virus. This needle biopsy taken from the liver of a mononucleosis patient shows infiltration of the liver parenchyma.

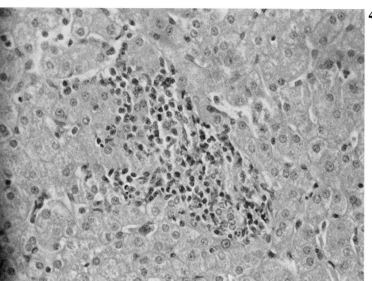

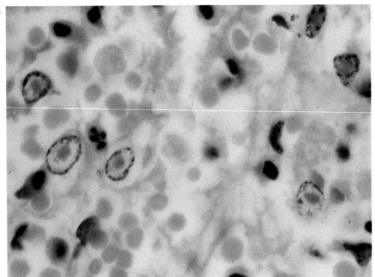

416 Herpes infection of the suprarenal gland. Herpes infections in immunocompromised patients can spread through all organs. The infection shown here shows the typical characteristic Cowdry type intranuclear inclusion bodies.

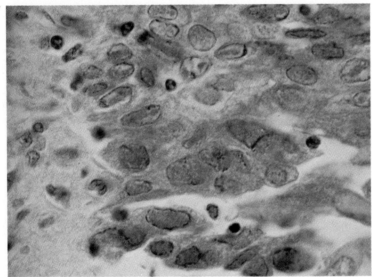

417 Herpes infection of the oesophagus. Herpes infections of the oesophagus mucous membrane are much more common than generally thought. In the bottom of the eroded lesions cells are found with typical herpes inclusions in the nucleus.

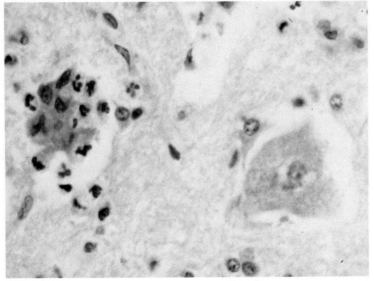

418 Poliomyelitis. Poliomyelitis is characterised by paralysis due to degeneration of motor neurons. At left a normal motor neuron and at right a motor neuron damaged by infection with poliovirus. The cell is surrounded with phagocytic cells which clean up the site. This phenomenon is called neuronophagia.

419 Tickborne encephalitis. Tickborne encephalitis is an arbovirus infection with a high rate of involvement of the central nervous system. Irreversible paralysis is due to destruction of motor neurons. In sections of the central nervous system neuronophagia can be observed.

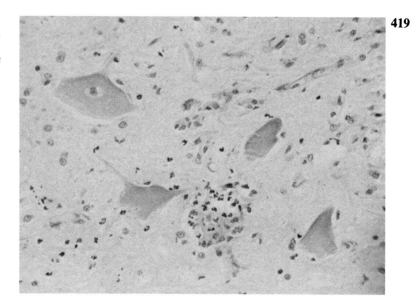

420 Tickborne encephalitis. The destruction of nerve cells can be observed in this section of the cerebellum in which a high percentage of Purkinje cells have disappeared.

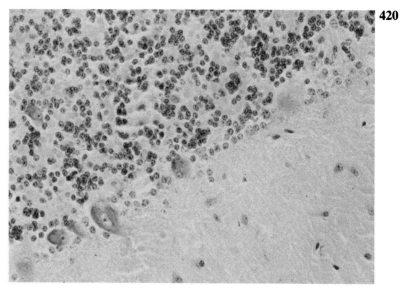

421 Rabies. Rabies virus infections are diagnosed at autopsy from sections of the hippocampus in which Negri bodies, pathognomonic for rabies virus, can be found.

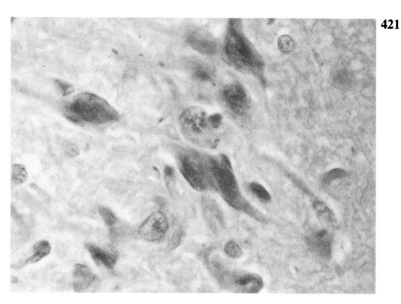

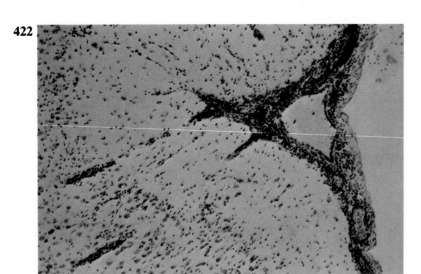

422 Herpesvirus, meningeal involvement. Herpesvirus can give rise to a meningoencephalitis in man and experimental animals. Here cellular infiltrations of the meningi in an experimentally infected rabbit are shown. The cellular infiltrations can be seen to progress along the blood vessels into the nervous tissue.

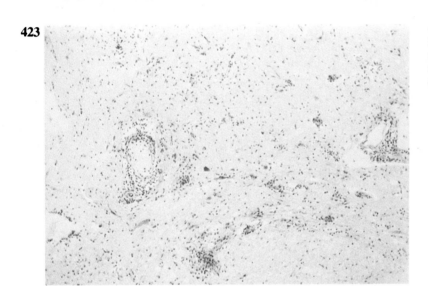

423 Poliomyelitis, perivascular infiltrates. Perivascular infiltrates are always present in acute infection of the central nervous system. Sometimes the phenomenom is called vascular cuffing.

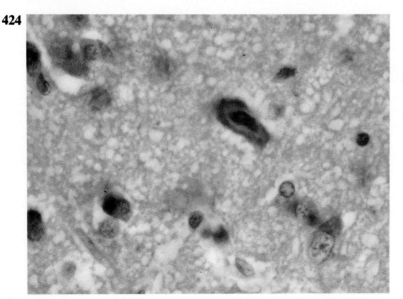

424 Herpes encephalitis. In this section of a case of herpes encephalitis a neuron containing a Cowdry type A inclusion can be seen. In biopsies these cells are considered characteristic for the disease and are detected by fluorescent microscopy or by staining.

425 Herpes in a ganglion gasseri cell. The herpesviruses type 1 and 2 and varicella zoster virus become latent in the cells of sensory ganglia. From here they can be reactivated and give rise to manifest infections. The nucleus of a neuron in this section of the ganglion gasseri shows typical inclusion bodies.

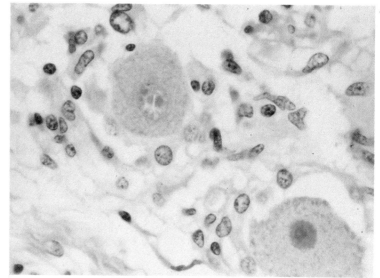

426 Cytomegalovirus of the brain. Cytomegalovirus can infect the brain, especially in the prenatal period and in young infants, laying the base for later school failure. In this section of the brain a congenital case of cytomegalo infection a typical 'owl's eye' can be seen.

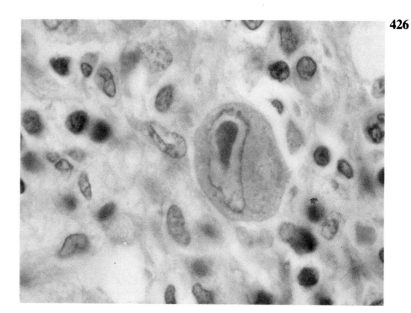

427 Lung in congenital cytomegalovirus infection. In congenital cytomegalovirus infections the liver, spleen, kidneys, pancreas, salivary glands and the lungs are involved. In the middle of this section of a congenitally infected lung a typical epithelial cell with an inclusion can be seen. There is a diffuse inflammation.

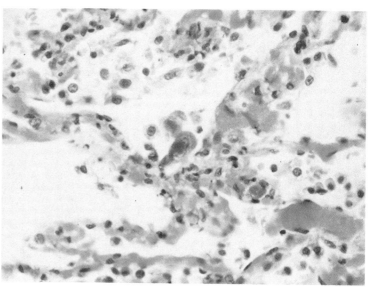

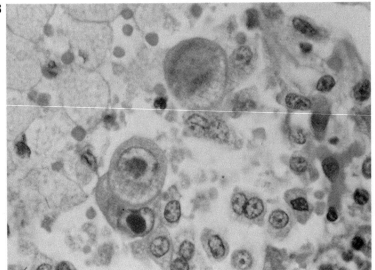

428 Cytomegalovirus infected lung in immunosuppression. Cytomegalovirus infection is a frequent complication seen in patients treated with immunosuppressive drugs. Biopsies of the lung show the typical enlarged cells with nuclear inclusions. The diagnosis can be confirmed by culture of the virus and by immunofluorescence.

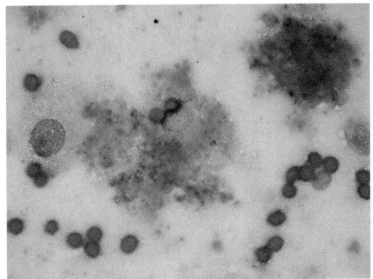

429 Pneumocystis carinii. Lung complications in immunosuppressed patients, often ascribed to cytomegalovirus, may be due to the parasite pneumocystis carinii which can be seen in sections or impression smears of the biopsy. The small round cysts can be seen in the middle of the picture.

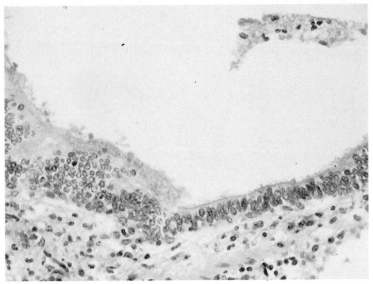

430 Giant cell pneumonia in respiratory syncytial infection. Respiratory syncytial virus is a frequent infection in young children. The virus is characterised by the ability to induce syncytia, giant cells. The giant cell (seen here in a fatal case) consists of hundreds of cells.

431 Giant cell pneumonia in measles. Measles infections give rise to formation of giant cells which can be found in nasal secretions and urine sediments. Measles pneumonia is characterised by giant cells, the nuclei of which have intranuclear inclusions.

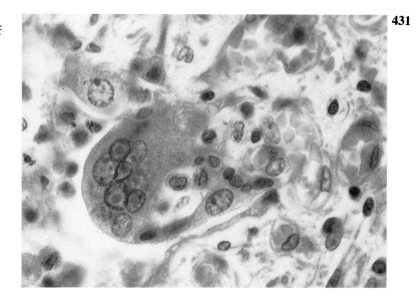

432 Normal tracheal epithelium. The mucosa is bordered by ciliated and goblet cells.

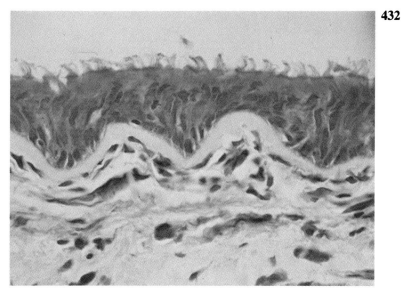

433 Epithelial lesions in the trachea in influenza. Typical picture of the patchy desquamation of the columnar eipthelium which occurs in uncomplicated influenza. At some spots in trachea and bronchi the destruction of the epithelium is complete, leaving the underlying tissue unprotected against bacterial invasion.

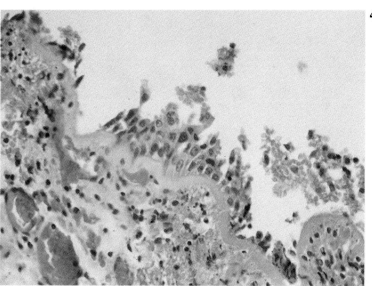

434 Secondary infection of the bronchial tree in influenza. At places where the epithelium is destroyed secondary infection has taken place. Note the dark patches caused by bacterial proliferation. This is a very serious condition.

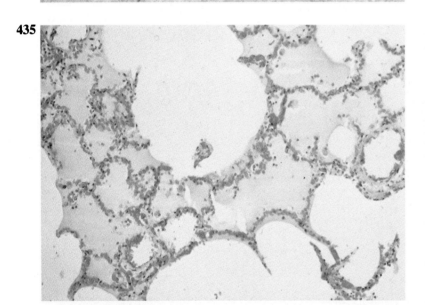

435 Fatal primary influenza pneumonia. After necrosis of the bronchial and bronchiolar epithelium has taken place, the alveoli are distended with haemorrhage and non-purulent oedema fluid. This extensive accumulation of fluid in the lung tissues leads to hypoxia. People with diminished respiratory function are most prone to this condition and should be protected by vaccination.

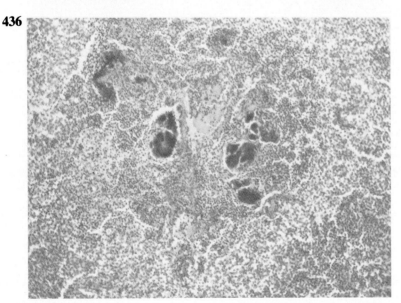

436 Secondary infection in influenza. Fatal influenza pneumonia complicated by staphylococcal infection. The foci of staphylococci can be seen as dark patches in the extensive cellular exudate. Resorbtion of staphylococcal toxin can lead to death within a few hours.

437 Varicella pneumonia. Varicella pneumonia can occur as a complication, particularly in adults with chicken pox. The virus infection causes rounding and desquamation of alveolar cells.

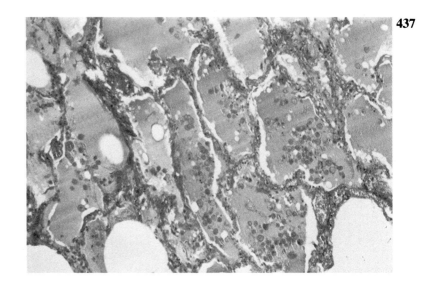

438 Varicella vesicle. The eruption of varicella consists of vesicles which appear within 2-4 days. The vesicle is formed by lysis of the cells of the epidermis and transudation of fluid. It can be observed in the picture that the genesis of the vesicle is multilocular.

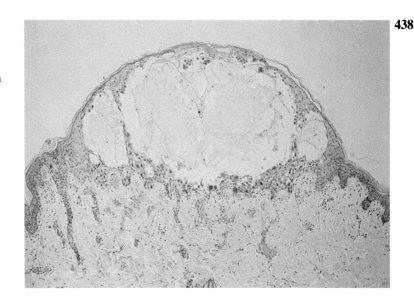

439 Varicella lesion. The bottom of a newly formed varicella lesion shows mild exudation, giant cells and other cells with intranuclear inclusions. As these cells contain a high titre of chickenpox virus, scrapings of the vesicle bottom are the best material for virus isolation.

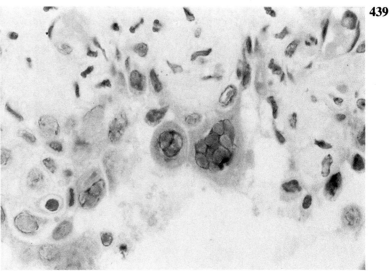

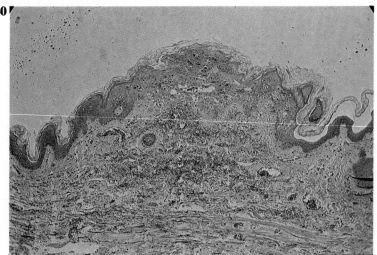

440 Haemorrhagic varicella.
Haemorrhagic varicella is a rare condition in which the vesicles show a haemorrhagic appearance. The microphoto of the cut through a developing lesion shows extravasation of blood in the underlying vessels. Haemorrhagic varicella is a rare condition mostly related to impairment of the immunological defence mechanisms of the host. Mostly the condition is severe, with symptoms like nose-bleeding, skin haematoma, vomiting of blood and high fever.

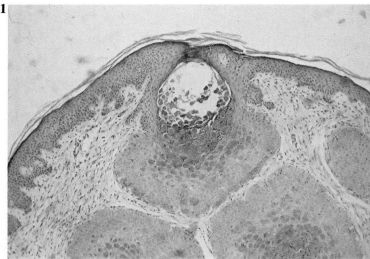

441 Molluscum contagiosum skin lesion.
Molluscum contagiosum is a small umbilicated skin tumour caused by a poxvirus. The lesion is a chronic proliferative process leading to the formation of a papule with a central opening showing a whitish mass. The infected cells contain a cytoplasmic inclusion body; the surrounding non-infected cells are stimulated to proliferate.

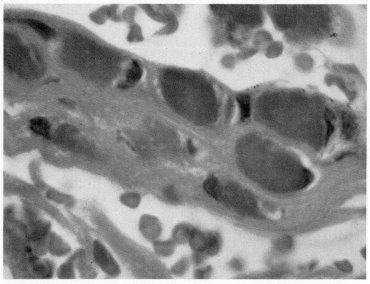

442 Molluscum bodies. The virus belongs to the pox group and multiplies in the cytoplasm of the cells, giving rise to typical eosinophilic inclusions called molluscum bodies. The nucleus of the cell is pushed aside.

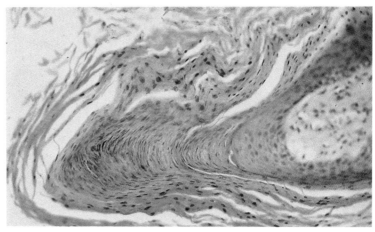

443 Common wart. The papovavirus which causes warts replicates in the skin cells causing a proliferative process. The horny layer formed by the infected cells is abnormal in appearance and desquamates with difficulty, giving rise to the typical appearance of a wart.

DIRECT TECHNIQUES FOR DETECTING VIRUS IN SPECIMENS

Viruses, viral antigens and viral nucleic acids can be demonstrated directly in specimens.
The following methods are available:

1 Direct staining of large viruses and viewing by the light microscope

2 Electron microscopy for morphological evidence of the presence of virus particles

3 Immunological techniques to detect viral antigens: immunofluorescence, immunoperoxidase, ELISA and RIA

4 Detection of viral nucleic acid in cells, even when there is no replicative activity

The direct techniques are used for rapid diagnosis of viral diseases. This can be necessary for differential diagnosis in the clinic, to prevent irrational therapy, for the surveillance and prevention of infectious diseases and in the near future as a basis for starting antiviral therapy.

A disadvantage of the direct and rapid methods is that one must be able to estimate as closely as possible what to look for, otherwise the search is time-consuming. Therefore the indirect methods are not becoming obsolete.

Viral isolation will complete the diagnosis and enable us to make a definitive identification of the virus associated with the current disease of the patient. Virus isolation is also an extremely sensitive technique as even one virus particle can yield a positive culture. Viruses can be isolated which are undetectable by other means.

ELECTRON MICROSCOPY

An electron microscope is an expensive instrument which may not be available in small laboratories. When available it can be of diagnostic value in a number of cases. Since several new human and animal pathogens were recognized for the first time by this technique, electron microscopy (E.M.) has been used as a routine screening for viruses, in case less intensive methods are not within reach.

The material to be investigated is suspended and centrifuged at low speed to get rid of debris. The supernatant can contain suffcient virus for E.M. preparations, otherwise the fluid is centrifuged at high speed to concentrate the virus in the pellet. Concentration can be done by adding antiserum to agglutinate the virus (immune electron microscopy, I.E.M.). Staining is done by the negative-staining method with phosphotungstic acid or uranyl acetate which make the unstained virus particles stand out. The diagnosis made by electron microscopy is morphological, on the appearance of the virus particles. Since viruses from a group have the same morphology, viruses cannot by typed with an electron microscope. Examples: all adenoviruses irrespective of their serotype or pathogenicity look alike. All herpesviruses, whether herpes simplex, varicella zoster, Epstein-Barr or cytomegalovirus, look exactly the same and E.M. diagnosis is no more precise than: 'herpestype virus'.

444 Grids for electron microscopy. Grids for electron microscopy are made of copper and must be coated with a carbon film on a formvar support before use. The grids are filed in special boxes and kept in an upright position. Watchmaker's tweezers are used to manipulate the grids.

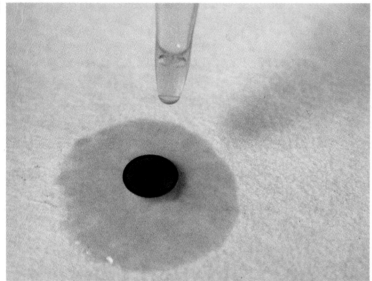

445 Bringing specimens to the grid. The clinical material, in the form of clear faecal extract, vesicle fluid or precipitated virus for I.E.M., is dropped on top of a grid placed on filter paper. The specimen is air-dried.

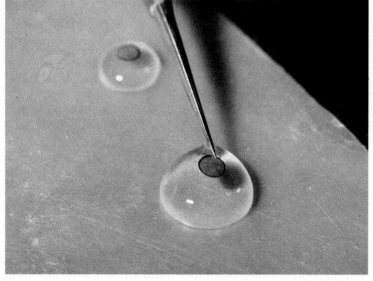

446 Negative staining. Drops of phosphotungstic acid (p.t.a.) are placed on a sheet of parafilm. The grids are floated in an upside down position on the p.t.a. After the proper staining time the grids are washed three times by floating them on drops of distilled water and subsequently air-dried by placing them on a filter-paper.

447 Adenovirus particles in the nucleus of a cell are arrayed in crystal like structures. Adenoviruses produce a cytopathic effect with characteristic intranuclear inclusions.

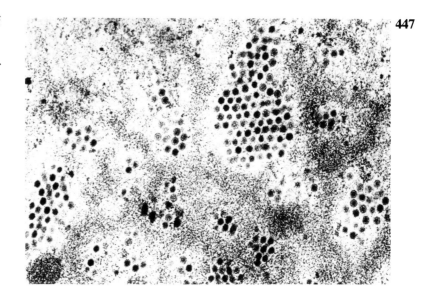

448 Purified adenovirus particles, in this case type 12. However all adenovirus particles look alike regardless of the type: icosahedral symmetry of the capsid which is 70 nm across. The number of capsomers is 252.

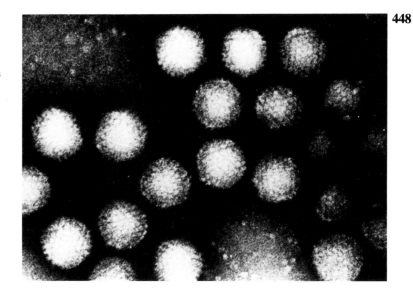

449 Adeno-associated virus as a contaminant of an adenovirus preparation. The adeno-associated viruses belong to the smallest viruses of vertebrates, with a single stranded DNA of 1.4×10^6 daltons. The A.A.V. is a defective virus only being able to replicate in the presence of adenovirus. Several types can be found in man and monkeys.

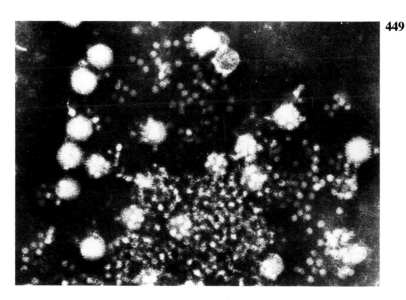

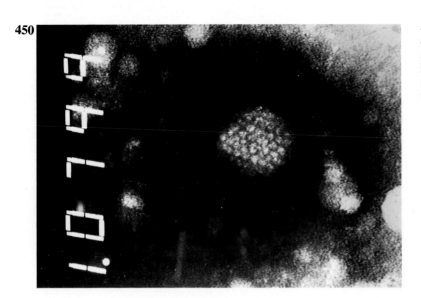

450 Adenovirus particle in a faecal sample of a child with diarrhoea. As the adenoviruses from the intestinal tract are hard to grow in tissue culture, electron microscopy is often the only way to make the diagnosis.

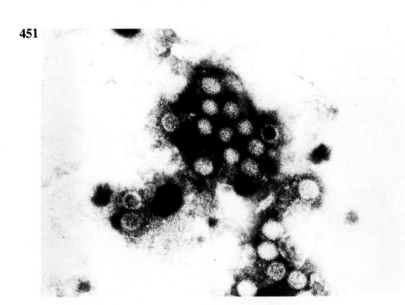

451 Rotavirus particles from a case of gastroenteritis. The virus is 70 nanometer in diameter and is hard to isolate in tissue culture. The virus was first detected by E.M. and the name of the virus was derived from the wheel-like appearance – which can be seen in some particles in this picture. As the virus can be readily detected in the faeces by ELISA, electron microscopy is little used for this diagnosis today.

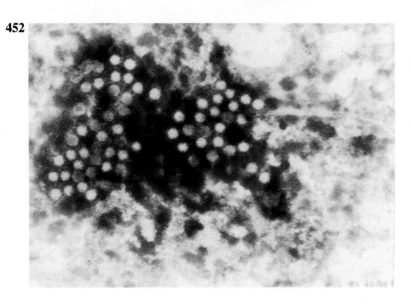

452 Norwalk-like virus particles. These 27 nm viruses can be detected in stool filtrates of patients with acute gastroenteritis. E.M. is the only means of diagnosis as the virus cannot be propagated in tissue culture.

453 Poliovirus in human faeces. The poliovirus belongs to the enteroviruses and has a diameter of 30 nm. All the viruses of the enterovirus group share the same morphological appearance and cannot be typed by E.M.

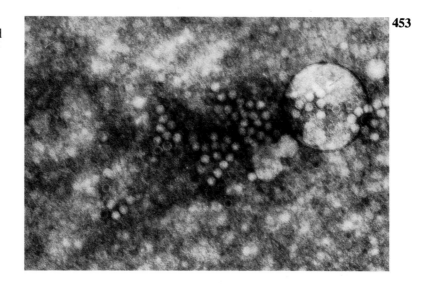

454 Human wart virus in the nucleus of an infected cell. Wart viruses are papilloma viruses having a diameter of 55 nanometer. These viruses can be readily extracted from warts and used as an antigen for serological reactions. Up to now propagation in tissue culture has been unsuccessful.

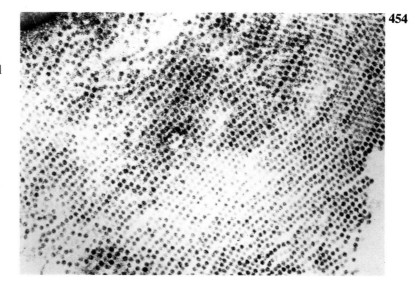

455 Thin cut of a cowpox lesion on the chorioallantoic membrane of a chick embryo. The ripe particles, 200 x 300 nm, are brick-shaped. Poxviruses are large enough to be rendered visible in the light microscope (see Figure 159).

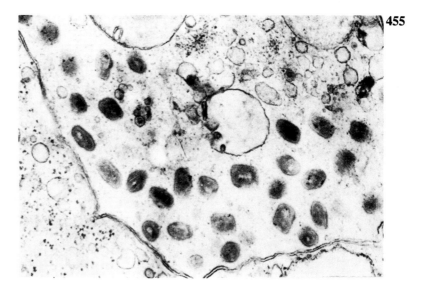

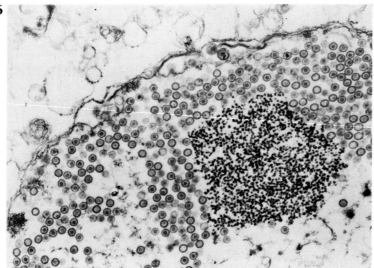

456 Herpesvirus type 1 particles.
Herpesvirus replicates in the nucleus, giving rise to a Cowdry type A inclusion body. The virus was inoculated into diploid human fibroblasts and the virus particles crowd near the membrane.

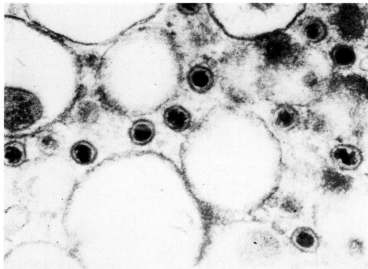

457 Herpesvirus particles in nucleoplasma.
In this thin section of an infected tissue culture cell typical herpes particles can be observed, measuring 180 nm in diameter. The capsid is icosahedral and composed of 192 capsomers. The virus particle is enveloped.

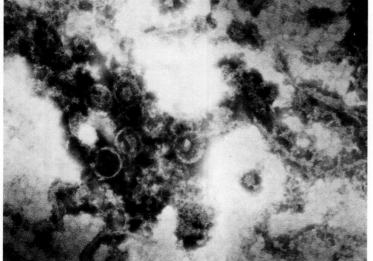

458 Smear from a fever blister negatively stained with phosphotungstic acid.
Enveloped particles can be seen in the debris from the vesicle. Under favourable circumstances a diagnosis – limited to establishing the presence of one of the herpesviruses – can be made in less than one hour.

459 Vaccinia virus replicating in a chick embryo cell. The viral nucleic acid is already produced and the subunits of the viral coat are now formed. The assembly of the virus in different stages can be observed. In the middle of the lower side a ripe particle can be seen.

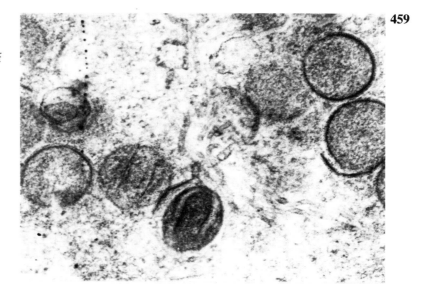

460 Electron micrograph of a purified preparation of SV40 virus, a simian virus belonging to the papova group of viruses. The particles are naked icosahedrons with a diameter of 45 nm and have 72 capsomers at the viral surface.

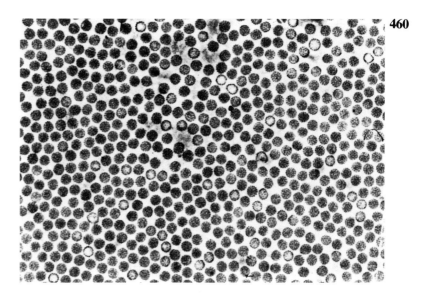

461 The DNA of SV40 virus is double-stranded and has a cyclic configuration as is revealed here by shadowing technique and high magnification. The MW of the DNA is only 3.5×10^6, the length of the DNA contour is 1·7 micrometer.

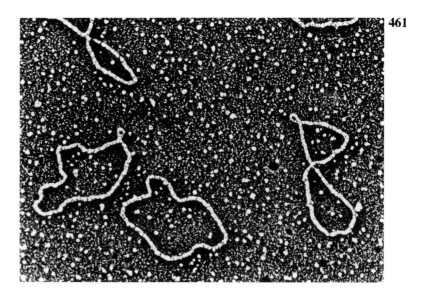

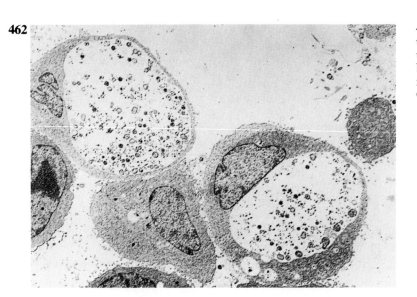

462 HeLa cells infected with Chlamydia trachomatis from a case of non-gonococcal urethritis. In the infected cells an inclusion body is formed in which the clamydia multiplies. Compare this E.M. with the Giemsa stained cells in Figure 335.

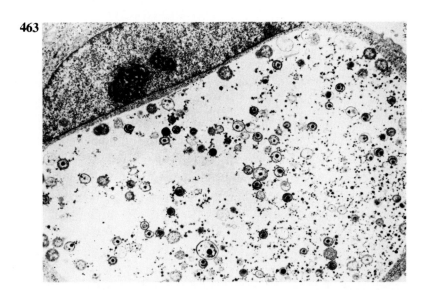

463 HeLa cells with chlamydial inclusion. The nucleus of the cell is in the top left corner. The inclusion contains the microorganism in different stages of development. The smaller ones with the black specks are elementary bodies, the larger near the nucleus the reticulate bodies.

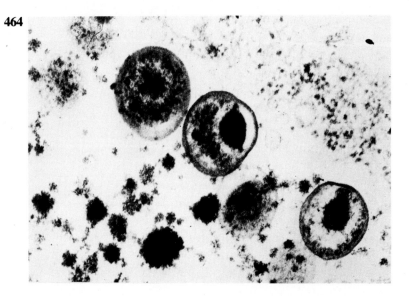

464 Elementary bodies of Chlamydia trachomatis. In the 500 nanometer bodies, at high magnification, ribosomes can be observed. Real viruses do not contain ribosomes. The elementary bodies shown are the infectious form of the chlamydiae.

IMMUNOFLUORESCENCE MICROSCOPY

Immunofluorescence (I.F.) microscopy is now commonly used to detect virus antigens in patient's samples as a method for rapid diagnosis, as a means of early detection of virus replication in tissue cultures used for isolation, and as a way to assay antibodies to viruses and their antigens.

The basis for I.F. is the labelling (conjugating) of antibodies with a fluorescent label, mostly fluoresceine isothiocyanate (F.I.T.C.), or in other cases tetramethyl rhodamine isothiocyanate. These labels light up in filtered ultraviolet light. Two methods are used, the direct method and the indirect method. The direct method uses labelled antibodies directed against the virus, allowing the staining to be done in one step with the advantage that fewer non-specific reactions occur. When monoclonal antibodies are used, non-specific staining is absent. The main drawback is that for every virus and for each antigen a conjugated antiserum has to be prepared.

The indirect method of immunofluorescence uses unlabelled antibodies to bind to the virus antigens and uses conjugated antibodies directed to the antiserum gammaglobulin to detect the virus-antibody complex.

The main advantage is that any antiserum can be used, and that only one conjugate is necessary for the serum of each species. For example rabbit antisera can be produced for all members of a group of viruses. Only one conjugate directed against the rabbit gammaglobulin, in this case for example F.I.T.C. conjugated goat antirabbit gammaglobulin is sufficient for the detection of virus by immunofluorescence microscopy. Another advantage is the slightly higher sensitivity. Disadvantages are the longer procedure, the slide has to be treated twice, and some more non-specific staining.

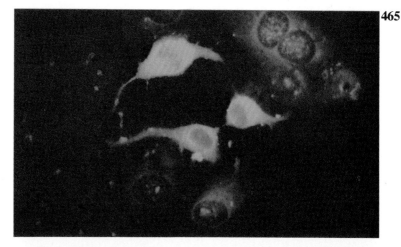

465 ECHO virus. ECHO virus type 8 was identified in infected cells by indirect immunofluorescence. The virus replicates in the cytoplasma which is showing positive fluorescence.

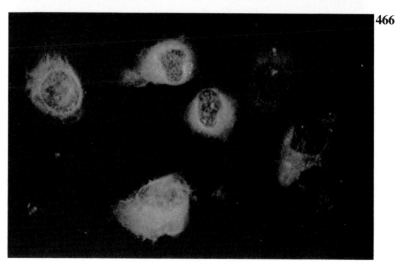

466 Polio virus. Polio virus type 1 in vero cells. Like all entero viruses polio virus gives a diffuse c.p.e. and replicates in the cytoplasma. In immunofluorescence diagnosis the localization of the fluorescence is of importance.

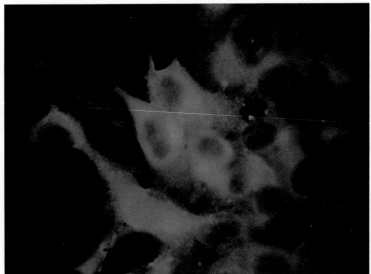

467 Vaccinia virus. Vaccinia virus in Sirc cells shows rounding of the cells and cytoplasmic fluorescence.

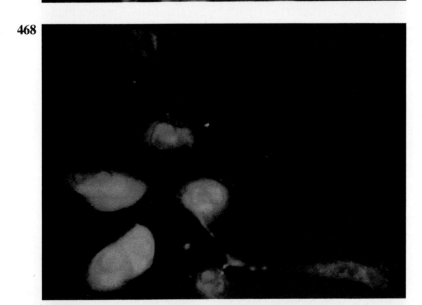

468 Cytomegalovirus. Cytomegalovirus in human diploid cells stained by human immune serum. The fluorescence is particularly strong in the nucleus where virus replication takes place. When cytomegalovirus fluorescence is done with cells inoculated at different times, different antigens, early and late, can be distinguished.

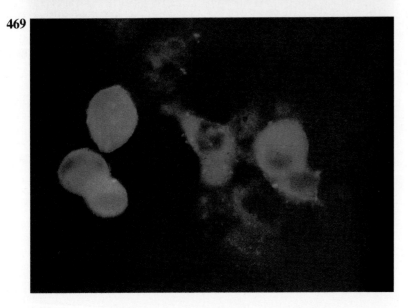

469 Adenovirus. Adenovirus type 5 in HeLa cells. Nuclear and cytoplasmatic fluorescence can be seen.

470 SV40 virus. SV40 on primary monkey kidney cells. The replication of the virus takes place in the nucleus. Nuclei which are not infected are dark or show only background fluorescence.

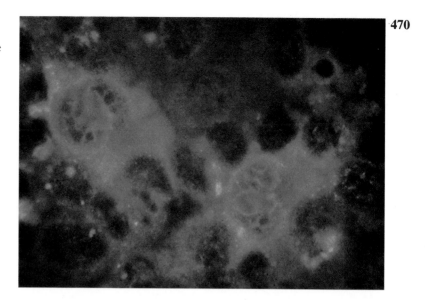

471 ECHO virus meningitis. Sayk sedimentation slides can be made from liquor cerebrospinalis. If the slides are made within half an hour after puncture, the cells will stretch beautifully and are well suited for immunofluorescence. Here an ECHO-infected cell with cytoplasmatic fluorescence is shown.

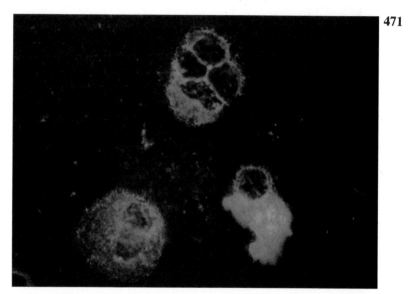

472 Varicella zoster in spinal fluid. Varicella zoster virus is difficult to grow from cerebrospinal fluid, but can be detected in many cases by fluorescence microscopy of the cells.

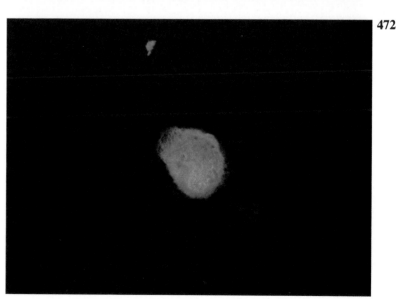

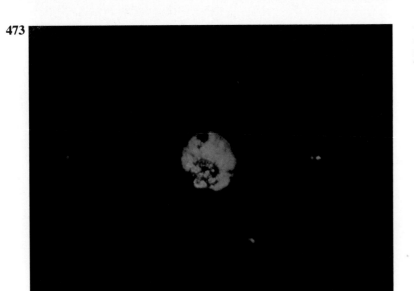

473 Measles meningitis. Measles virus in cerebrospinal fluid can be detected by fluorescence with an animal antiserum. The fluorescence is typically spotty.

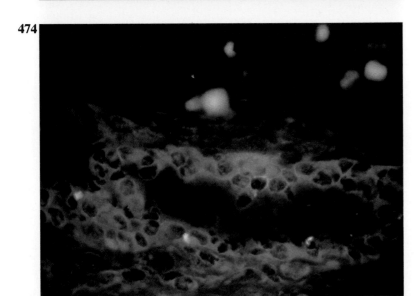

474 Rubella virus in an embryonic heart. The heart from a congenitally infected embryo was cut on a freeze-microtone. Staining with an animal antiserum revealed the virus multiplying in the cells.

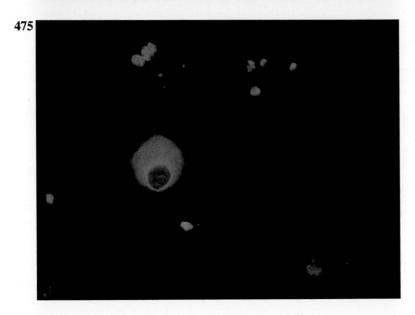

475 Rubella virus in amniotic fluid. Amniotic fluid was drawn from the child of a pregnant woman infected with rubella. The infected cells show cytoplasmic fluorescence, the non-infected are dark. The slide was made in a Sayk sedimentation chamber.

476 Rubella virus in urine. A urine sediment from a newborn child with congenital rubella. A number of cells show positive immunofluorescence.

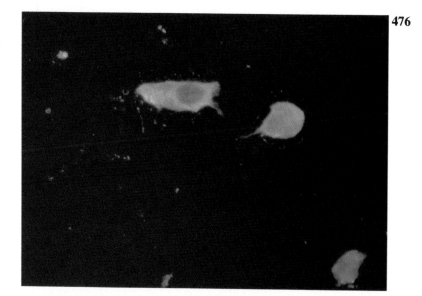

477 Reovirus inclusion body. Reoviruses replicate in the cytoplasm and form large eosinophilic inclusions. The inclusion consists of virus material as is shown by fluorescence with a specific antiserum.

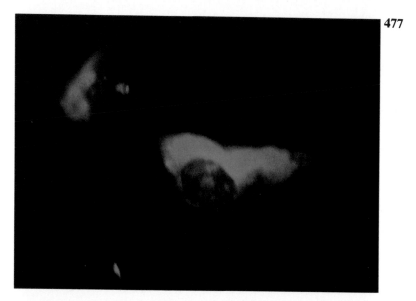

478 Herpes simplex virus typing in human kidney cells. Material from a genital lesion was inoculated in cell culture. The second day the cell sheet was treated with antiserum against herpes type 2 and a rhodamine conjugated gammaglobulin. The primary lesions, which were not yet visible as plaques, are stained intensely red.

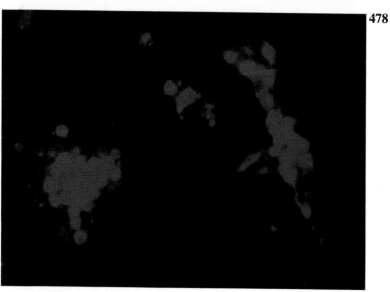

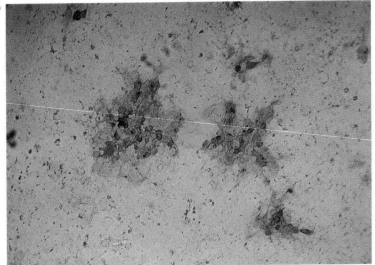

479 Immunoperoxidase staining of herpes simplex virus type 1. Herpes virus type 1 was inoculated in vero cells and 24 hours afterwards the culture was fixed, treated with an antiserum and a peroxidase conjugated second antibody, resulting in an insoluble brown product staining the virus-containing cells. The primary foci of virus multiplication are visible at low magnification.

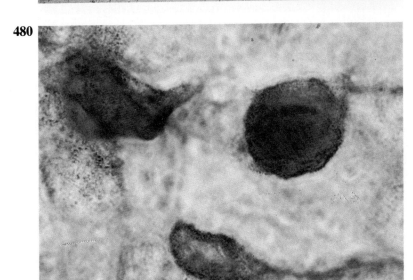

480 Immunoperoxidase stained cells. Cells infected with herpes type 1 showing cytopathic effect were treated with an immunoperoxidase stain. Brown deposits are formed where viral antigen is present.

HYBRIDIZATION TECHNIQUES

Hybridization techniques have recently been developed and are very useful for viral research, but as an occasional resource for difficult diagnostic cases, not for routine diagnostic work.

The hybridization technique is based on the fact that single-stranded nucleic acid will re-anneal in a characteristic way with complementary strands. Recombinant DNA techniques are used to produce complementary DNA strands by production in bacterial plasmids. The nucleic acid strands, to be used as probes for the detection of viral nucleic acid, are labelled with fluorescent dyes, biotin, or isotopes.

The specific viral DNA sequences are extracted from the infected cell, absorbed on nitrocellulose membranes and hybridized with a radio-isotope-labelled DNA probe. The hybrids are detected by autoradiography. Positive spot hybridization manifests as black spots on the photographic emulsion.

In situ hybridization is used when the viral DNA has to be detected inside specific cells or tissues. When a fluorescent label is used the hybrids show up in ultraviolet light. A better relation to the structures can be had when a radiolabel is used with autoradiography and staining of the cells.

RNA complementary to DNA can be prepared by using RNA polymerase. The cRNA can be hybridized with the DNA. A rabbit serum can be prepared against this RNA-DNA hybrid and the antibodies fluorescein-labelled. *In situ* hybridization can now be done by hybridizing the DNA with cRNA and the hybrids can be detected by fluorescent microscopy.

RESTRICTION ENDONUCLEASE ANALYSIS OF VIRAL DNA

The endonucleases attack the nucleic acid chain at certain points, in a non-random way. The fragments are characteristic for a certain DNA. The nucleic acid of related virus strains can be compared when the restriction fragments are subjected to poly-acryl-amide gel electrophoresis. The bands can be stained with a fluorescent dye and photographed in ultraviolet light, or can be detected by autoradiography. The genetic relationship of viruses can be seen from localization of the bands.

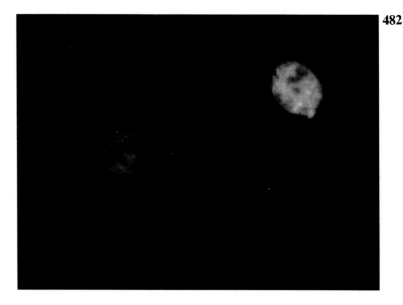

481 Detection of viral antigens by the DNA-RNA hybridization technique.

   1 Cell with viral antigen to be detected
   2 Hydrolysation of the cell
   3 RNA complementary to the viral DNA is added, *in situ* hybridization occurs
   4 Rabbit antibodies directed to DNA-RNA and FITC conjugated are added
   5 Inspection of the cells with a fluorescent microscope reveals the site where the viral DNA is present

482 **The presence of BK papovavirus DNA in the nucleus of the infected cell** is revealed by the DNA-RNA hybridization and staining with rabbit anti-DNA-RNA conjugated with F.I.T.C. The BK virus DNA stains green, the cell nucleus DNA is not stained.

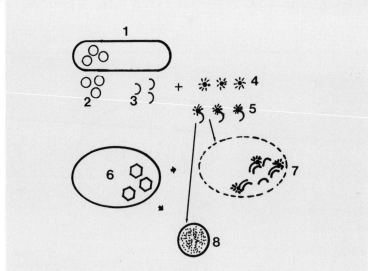

483 In situ and spot hybridization with labelled DNA/DNA hybrids

1 E coli with plasmids which contain viral DNA fragments (recombinant DNA technique)
2,3 Plasmids are purified and cut open
4,5 Conjugation of the DNA fragment with a fluorescent or radioactive label
6 Cell containing virus
7 Denatured DNA is hybridized in situ with the labelled DNA
8 Or the cellular and viral DNA are put on a nitrocellulose sheet and spot hybridization is done with the labelled DNA – an autoradiogram registers the amount of radioactivity present.

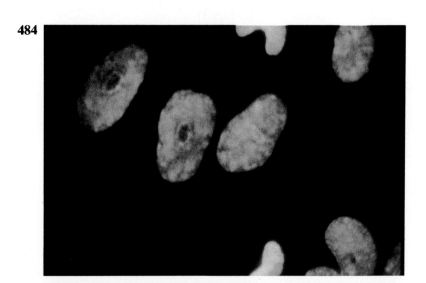

484 Human diploid fibroblasts infected with cytomegalovirus which replicates in the cell nucleus. The slide is stained with dapi (4:6 diamidino-2-phenylindol) a fluorochrome that stains DNA blue in ultraviolet light. The dapi stain reveals the total amount of DNA present.

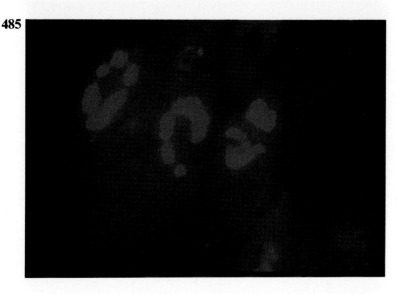

485 The same cells as shown in the previous picture. In this case the specific cytomegalovirus DNA is stained bright red after hybridization of the viral DNA with DNA labelled with AAF (acetyl amin fluorine).

486 In situ hybridization of a virus-infected cell in a paraffin section of a cmv infected lung. The DNA for the hybridization was rendered radioactive by labelling with an isotope. An autoradiogram was made and the infected cell can be recognized by the dense pattern of silver grains.

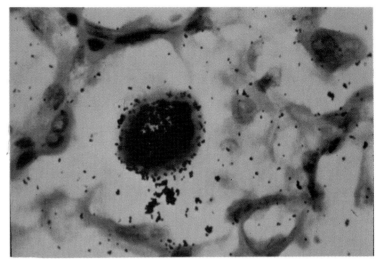

487 Spot hybridization of DNA on nitrocellulose paper. DNA was extracted from infected cells and absorbed on nitrocellulose. After hybridization with isotope labelled DNA an autoradiogram was made. Dark spots are cell extracts positive for viral DNA; the light spots are negative.

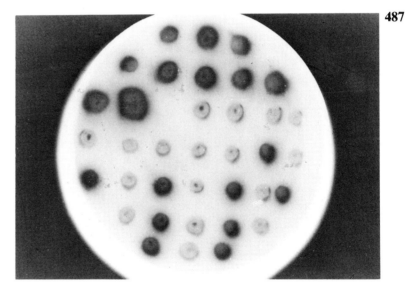

488 Restriction endonuclease analysis of DNA. Three adenovirus strains were treated with different restriction enzymes and subjected to electrophoresis, including references and untreated virus. The genetic relationship between the strains can thus be established.

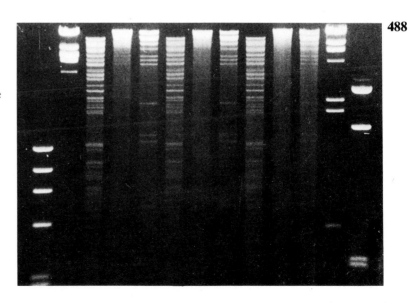

Serological Tests

Serological tests are based on the reaction between antibodies and antigens. They are used either to detect antibodies or to detect antigens.

Serological techniques play a big role in the management of virus infections as they can be used to make the diagnosis even when virus isolation is not or no longer possible. The basis of serology is the specificity of the reaction and the circumstance that it takes time for the individual to build up a sufficient level of antibodies. In most cases 2 to 3 samples of serum are taken with one or two weeks between them, it being necessary to get at least a four-fold rise in titre as evidence of a current infection. This scheme is not fixed for all tests as different kinds of antibodies develop at different times.

IgM class antibodies are – generally speaking – early antibodies and tests to demonstrate them are used for rapid diagnosis. The titre of the serological reaction is not to be used as an indication of an existing illness. There is so much individual variation that no normal values can be given. On the other hand it is not true that the height of the titre attained in a test indicates that this test is more accurate than any other test.

Which type of serological reaction is used in a case depends on what is to be done with the results. Specific reactions, like the neutralization test, are useful when the degree of immunity is assayed. In other cases it is advantageous to use tests which are group specific. For example all adenoviruses share a common complement-fixing antigen. Thus any adenovirus infection, independent of the type, can be diagnosed by a four-fold rise in the complement-fixation test. Cross-reactions can be misleading, for example in arbovirus tests, and in these cases the most specific tests must be employed. A source of false-positive reactions are non-specific reacting substances in the serum, which must be removed before the serological test is performed.

Blood for serological purposes must be drawn in a sterile, absolutely dry, tube. The modern vacuum systems are well suited. Take the blood prior to a meal to avoid getting lipaemic serum. Avoid bacterial contamination and haemolysis.

Serum can be stored for a few days at $+4°C$, up to 5 years at $-20°C$ to $-30°C$ and when lyophilized indefinitely at $+4°C$.

Sera to be used for antigen detection, immunofluorescence, enzyme linked immuno sorbent assay (ELISA), radio immuno assay (RIA), precipitation, etc., should be high titred and type-specific. Monoclonal antisera have a narrow band of specificity; sometimes better results are had when sera from related clones are mixed. The fluorescein iso-thio-cyanate (FITC) and enzyme conjugated antigammaglobulin sera are of animal origin and may contain non-intended antibodies against viruses, thus interfering with the reactions. For example, goat serum may contain antibodies against enteroviruses which will give rise to false positive reactions in the fluorescent assay test (FAT) in serology.

PRETREATMENT OF SERA USED IN SEROLOGICAL TESTS

As all sera contain non-specific reacting substances that influence the results of the serological reactions with antibodies, measures must be taken to remove these interfering substances. When a complement fixation has to be performed all complement has to be removed from the serum. This is done by heating the prediluted serum at 56°C for 30 minutes or at 60°C for 20 minutes. A number of erythrocytes is agglutinated by normal human serum. Well known examples are sheep erythrocytes and goose erythrocytes. If non-species red blood cells are to be used for a serological reaction the non-specific agglutinins must be removed by absorbing the serum with the red blood cells, otherwise the serum will agglutinate the erythrocytes and the results of the reaction are incorrect.

When human erythrocytes are used in serological reactions blood-group O and rhesus negative cells are employed which are not agglutinated by other blood group agglutinins. Several substances in normal sera can react with viruses in a non-specific, that is non-immune, way, mimicking a serological reaction. The non-specific reacting substances can be removed by treatment with neuramidase, potassium periodate, trypsin, manganese chlorideheparin or kaolin, depending on the reaction to be performed. In some cases the determination of the antibody class is of particular importance. Mostly IgG and IgM antibodies are to be titrated separately. The oldest and simplest method is separation of both antibodies by centrifugation in a sucrose density gradient. The gradient is made by layering different densities of saccharose over each other. The serum mixed with F.I.T.C. is layered on top. By centrifugation in a swing out rotor at 35/40.000 r.p.m. for 8 hours IgG and IgM separate.

If the serum of the patient contains rheumatoid factor this factor can bind to the IgG causing the complex to be layered in the IgM band leading to a false diagnosis. The same mistake can be made when the IgM separation is performed by gel filtration. The rheumatoid factor must be removed by adsorption to IgG insolubilized at 73°C. Removing IgG with anti Fc will give IgM without rheumatoid factor.

489 The microtitre plate. The microtitre system is based on the use of 96 preformed depressions in a translucent plastic plate. These cups are used instead of the classical test tubes. The plates are standardized and the place of each cup can be indicated by number and letter.

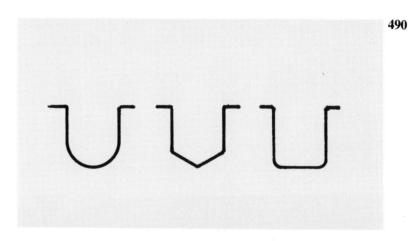

490 Profiles of microtitre cups. The microtitre cups are used in three different profiles, the flat-bottom type, the round-bottom (U) type and the pointed (V) type.

491 U type microtitre cups. The U type cups are the descendants of the original test tubes. The round bottom allows sediments, mostly red blood cells, to settle at the deepest point, which enhances the reading of the test. These cups are mostly used for complement-fixation tests.

492 V type microtitre cups. The V type microtitre cups have a bottom with an angle. This allows sediments to collect as a sharp point in the slant of the bottom, allowing a better readout for haemagglutination patterns.

493 Flat bottom cups. Microtitre plates with flat bottom cups are purpose-made. In serology they are employed for the ELISA test for which a specific quality is needed. When these plates are used for tissue culture the bottom must have all the properties needed for anchorage of the cells.

494 Pipette for microtitre use. The volumes used for pipetting in microtitre techniques are standardized to 25, 50 and 100 microlitre. Pipetting can be done by using individual pipettes like the one shown here.

495 Disposable tip pipettes.
Modern single or multiple drop pipettes employ disposable tips of non-wettable plastic. The fluid is drawn and blown out by air displacement. The pipette tips are colour-coded to indicate the maximum capacity.

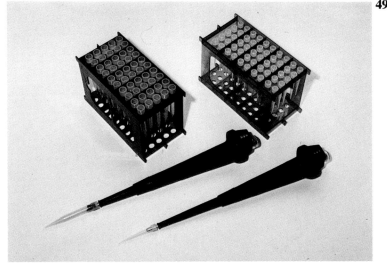

496 Multiple channel pipettes.
If the workload is not enough to use a pipetting and diluting machine multiple channel pipettes can be used to fill the microtitre wells and to make the dilutions.

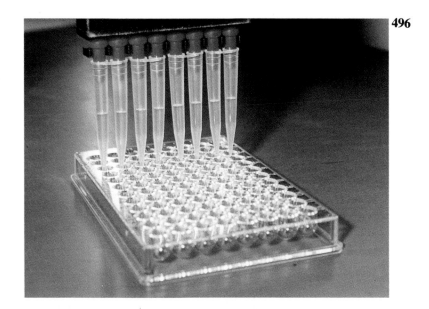

497 Pipetting machine. To ensure a high production in a service laboratory cell, routine handlings must be mechanized. The machine shown here has a pipetting head which is able to fill all 96 wells of the microtitre plate at once with a fixed amount of dilutant buffer, complement, antigen or haemolytic system and so on.

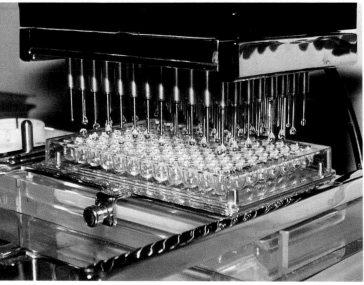

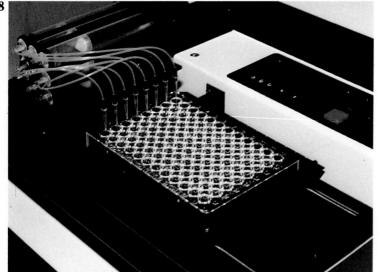

498 Multichannel pipetting machine. The pipetting machine can be adjusted to dispense a predetermined amount of liquid to a row of 8 wells at a time. The microplate is moved mechanically under the tips until all wells are filled.

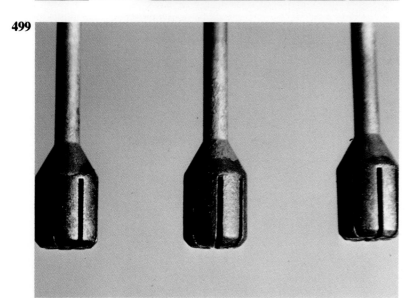

499 Diluter tips. Dilutions are made in the microplate system by carrying over and mixing 25 microlitre amounts with non-corrosive metal tips. After each use the tips are cleaned by rinsing in buffer and blotting. After 20 dilutions, or at the end of the job, the diluters are heated in a flame.

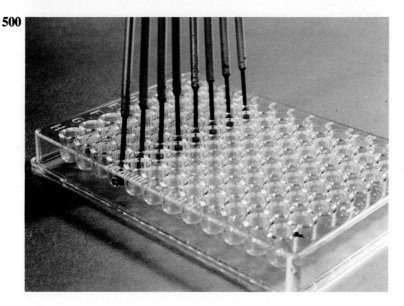

500 Hand-operated diluters. Small numbers of sera can be diluted by hand using these microdiluters mounted on long tapered handles. Experienced workers can handle 8 tips at a time.

501 Testing microdiluters. The microdiluters are regularly tested by filling them and blotting them on a microdiluter tester. The right amount of 0.025 ml will wet exactly a printed circle. If not, the diluter must be discarded.

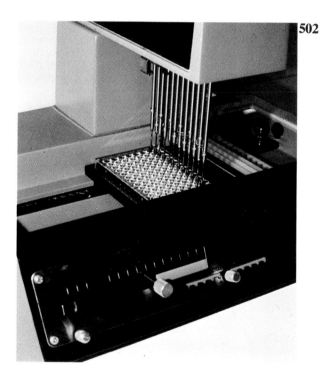

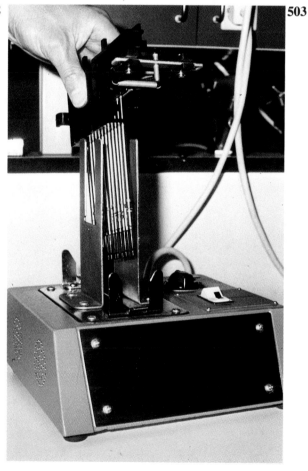

502 Diluting equipment. Mechanically operated diluters, many of them completely programmable, will speed up the diluting process. The machine allows diluting to take place 8 ways or 12 ways.

503 Heating the diluters. The diluters in the cassette are rinsed in buffer and blotted after each dilution and must be heated after 20 dilutions. This heating can be done in a gas flame or in an automatic electric heating unit like the one pictured here.

504 Microplate covered with lid. The contents of the wells can be protected from evaporation and contamination by covering them by a lid. In some cases the lid has a notch so that it can be used in one way only, preventing mixing up.

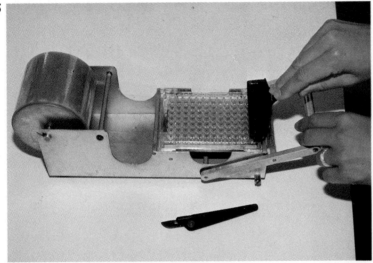

505 Microtitre plate sealed with tape. Sealing tape can be had pre-cut or on rolls. The sealing tape is pressed firmly with a roller. Sealing is done before centrifugation and when prolonged storage is desirable.

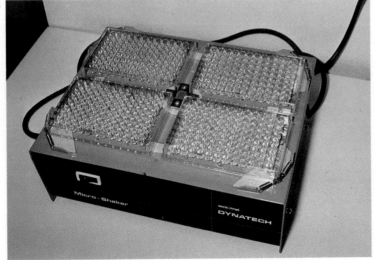

506 Plate shaker. As the contents of the wells cannot be mixed, as is done in tubes by sucking up and down a pipette, mixing must be done by shaking the plate. Shakers can accommodate 1, 2 or 4 plates.

507 Centrifugation of microtitre plates (1). To speed up operations and to get a clearer reading microtitre plates can be centrifuged. This can be done in the large refrigerated centrifuges like those of Christ. Special adapters are made in which the plates are placed.

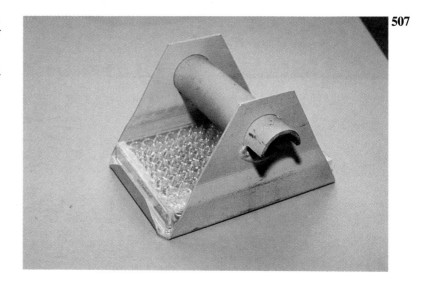

508 Centrifugation of microtitre plates (2). If cooling is not important the plates can be centrifuged in a small table centrifuge specially adapted for the purpose.

509 Reading the microplate. The easiest way to read a microtitre plate is to place it on top of a special plate reader mirror which can give either a normal size reflected image or an enlarged one.

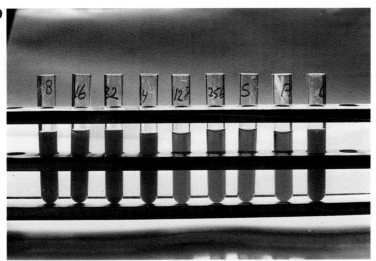

510 Demonstration of the complement-fixation test in tubes (1). A twofold dilution series of the serum is made, and controls are included without the serum (s), the antigen (a) and the complement (c). Dilutions 8, 16, 32 and 64 are positive. The complement is used up and the erythrocytes are not lysed. The controls are correct.

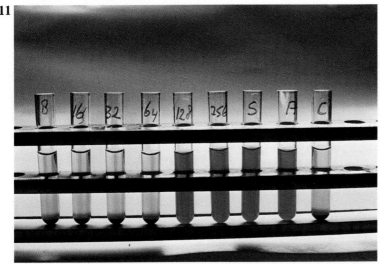

511 Demonstration of the complement-fixation test in tubes (2). When the complement-fixation test is placed in the refrigerator the erythrocytes in the positive tubes will form a sediment leaving a clear supernatant. The negative tubes, which are haemolysed, stay red.

512 The complement-fixation test in a microtitre plate. The complement-fixation test in microtitre plates is always refrigerated before reading so the erythrocytes form a sediment. Red means negative, a dot is positive and white are the wells which were not used. Sera 1–12 were diluted 1:8 to 1:256 to determine the complement-fixation test titre.

513 Report of the complement-fixation test in microplate. All serum controls are right. Serum 1, 2 and 9 are negative. Serum 7 and 8 are 1:8; serum 2, 10 and 11 are 1:16; serum 5 and 6 are 1:32: serum 4 is 1:64; serum 12 is 1:128.

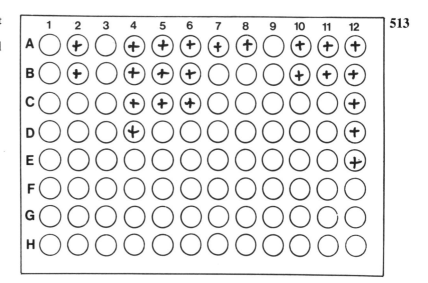

514 Complement-fixation test as a screening test. When large numbers of sera have to be tested it can be an advantage to use the complement-fixation test as a screening test. Each well contains a different serum in a single dilution (1:16) with antigen and complement added. Every serum shown positive has to be tested in dilutions with appropriate controls.

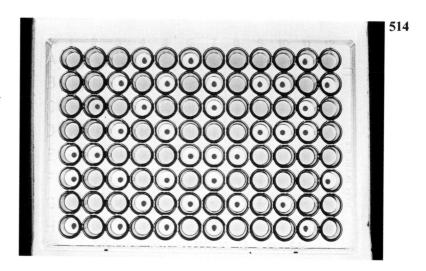

515 Reports of the screening test. Positive are sera 1 E, F, H; 2 E; 3 B, D, F, H; 4 A, C; and so on (+).
Serum 2 C gives a weak reaction as some haemolysis shows (±). Wells showing haemolysis contain negative sera (−).

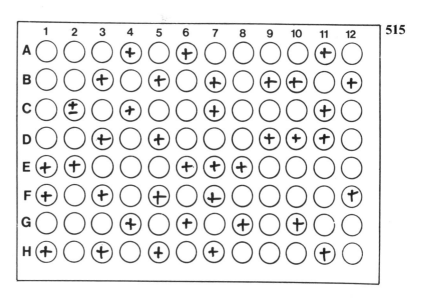

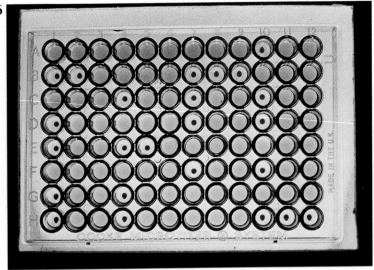

516 Complement-fixation test screening with controls. Complement-fixation test screening in three successive rows, first row serum + antigen + complement, second row serum + control antigen + complement, third row serum + saline + complement. This method gives the possibility to detect false-positive and anti-complementary sera.

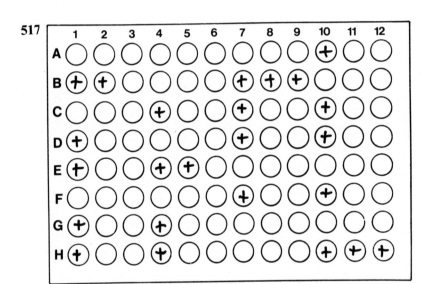

517 Reports of the screening test. Positive sera 1 D, E, G, H; 4 C, G, H; 7 C, D, F; 10 A, C, D, F. (+)

False-positive (reacting with the control antigen) are 1 B; 4 E. (F)

Anticomplementary are 7 B; 10 H (AC).

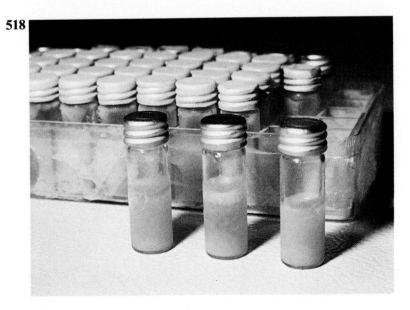

518 Aliquots of complement. Complement is guineapig serum made within a few hours of bleeding. The serum of a number of animals is pooled and filled into small bottles containing 1 to 1.5 millilitre. The aliquots are stored at −70°C. Wait 14 days before titrating the complement.

519 Titration of the complement. As the exact determination of the titre of the complement is important 8 dilutions are made of 2 aliquots. Starting at 1:10 up to 1:120. In this titration the dilution 1:50 (row 5) gives 100% haemolysis. The titre is 50, the working dilution containing 2 full units is 1:25.

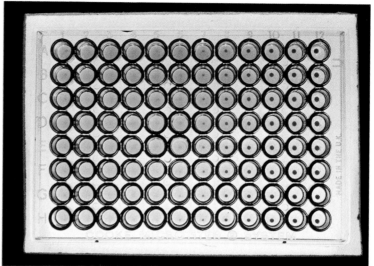

520 Titration of an antigen. Every antigen used in the complement-fixation test must be tested in a block-titration against a known positive serum. Rows A to G (horizontal) contain a serum dilution of 1:8 to 1:512, H is the serum control. Rows 1 to 12 (vertical) contain a dilution series of the antigen: 1:1 to 1:12.

At dilution 11 the antigen shows a sharp fall in activity. At dilutions 6, 7, 8, 9 and 10 the activity is constant. As 10 is the borderline dilution we use the antigen four dilutions stronger, that is at 1:7.

521 The immune adherence haemagglutination assay (IAHA). The immune adherence haemagglutination assay is a serological reaction in which the antibodies react with the antigen while the complement is added afterwards. The complement pathway is stopped by dithiothreitol and human O red blood cells are added. The antibody-antigen-complement complex adheres to the receptors on the erythrocytes causing them to agglutinate.

  1 Inactive serum at 56°C 30 min.
  2 Add antigen, 60 min. 37°C
  3 Add complement, 40 min. 37°C
  4 Add dithiothreitol; add O erythrocytes 120 min. 37°C
  5 Read the agglutination pattern

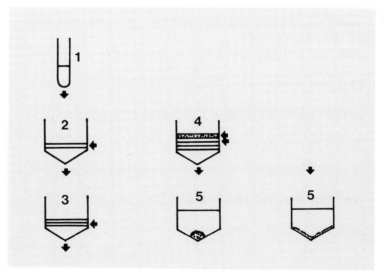

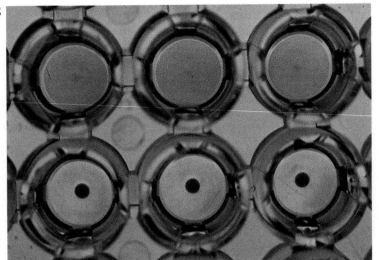

522 Agglutination pattern of the IAHA.
The positive reaction in the IAHA results in an agglutination with a granular appearance, the pattern being fine or somewhat coarse, depending on the nature of the antigen. In the negative wells the erythrocytes form a distinctive dot of sediment. False positive reactions can occur, making it necessary to use controls on all sera.

523 The IAHA in microtitre plates. V type plates are used. The sera are made into twofold dilution series in double rows of 12 wells. The first row is the IAHA; the second row the control.

The test can be read as soon as the red blood cells are sedimented and the haemagglutination pattern formed. Reading the test after it has stood overnight at 4°C is good practice.

The first serum reaches a titre of 4096, the control is non-reacting.

The second serum has a titre of 512, the control being negative.

The third serum gives a titre of 128 which is false-positive as the control reaches the same value.

The fourth serum is negative.

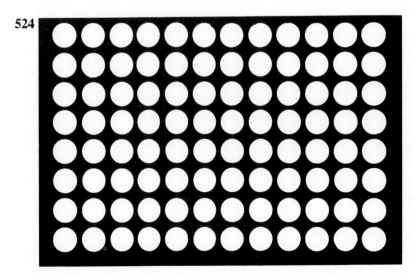

524 Microtitre antiglare shield. As the whole microplate is translucent, fine differences in the reaction are sometimes hard to read. In this case an antiglare shield is used which comprises a black plastic sheet with openings corresponding to the microtitre wells.

525 IAHA with glare shield. The same plate as in Figure 523 is shown here in position with an antiglåre shield, proving that a high reading contrast can be attained.

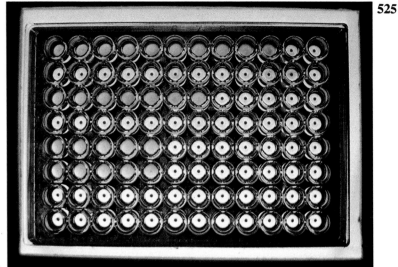

526 Antigen titration for the IAHA. The antigens used for the IAHA are the same as for the complement-fixation test but the optimal dilution is different. The block titration of the antigen is done in the same way as for the complement-fixation test. The highest dilution which gives optimal results is determined and a certain excess is used.

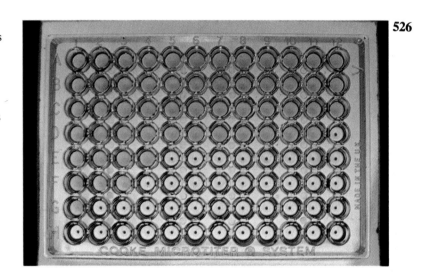

527 Haemagglutination.
Haemagglutination means agglutination of red blood cells. A number of viruses is able to adhere to the surface of erythrocytes, in most cases by an enzymatic reaction, sometimes by a lipid reaction. Viruses adhere to more than one erythrocyte, resulting in clumping together.

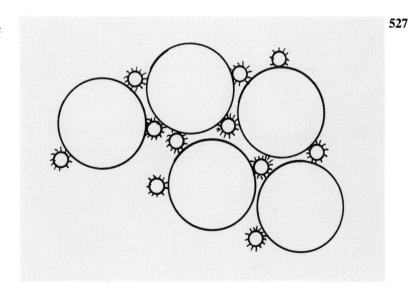

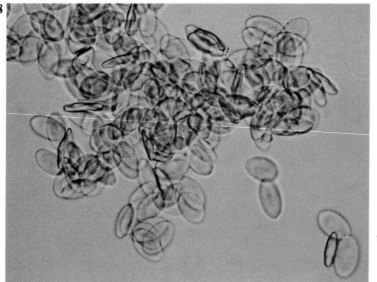

528 Agglutinated chicken erythrocytes. Chicken erythrocytes, which are oval and have a nucleus, agglutinated by influenza virus. The picture shows the forming of smaller and larger aggregates. In influenza the aggregates are stable at 4°C, at higher temperature the virus is eluted and the haemagglutination disappears.

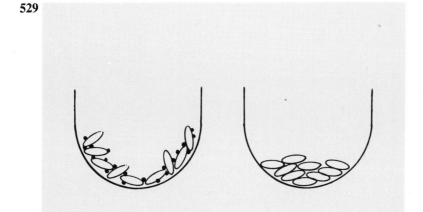

529 The haemagglutination pattern. A U or V plate is used. The non-agglutinated erythrocytes settle to the bottom of the well and slide down to the deepest point forming a smooth round heap of red cells. The agglutinated erythrocytes are not free to slide down to the bottom and form a distinct pattern.

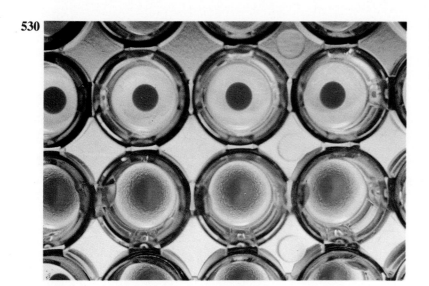

530 Negative and positive haemagglutination. In this V plate the negative haemagglutination is characterized by the sharp red dot. The agglutinated erythrocytes form a very distinctive pattern at the bottom.

531 Controlling the haemagglutination pattern. When the plate is put in upright position, at nearly 90° angle, the haemagglutinated erythrocytes will stick together. The red blood cells (r.b.c.) which are not agglutinated will flow freely. This trick is often used to control haemagglutination and haemagglutination inhibition when the virus does not give very distinct patterns. Such tests are read by observing the sticking together of red blood cells.

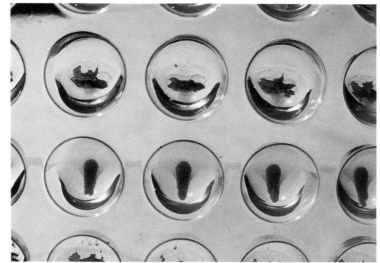

532 Absorbing sera with kaolin. In a number of reactions non-specific inhibitors in serum interfere with the reaction. To remove these lipoid-like substances the sera are absorbed with a suspension of kaolin. During the reaction time serum and kaolin must be thoroughly mixed. This is best done with a roller apparatus as shown here.

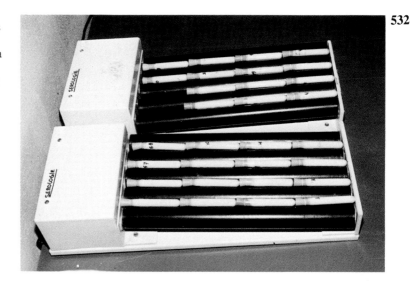

533 Haemagglutination test. The haemagglutination test is used to determine the titre of virus isolates. In this V type microtitre plate the batches A to H are diluted from left to right. The 12th row is the erythrocyte control to which no virus dilution is added.

All batches of influenza virus proved to be positive in the haemagglutination test. The highest titre (1024) was reached by batch C. The lowest titre (8) is batch A, the fourth tube giving a partial reaction.

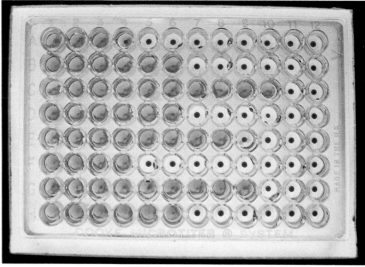

534 The haemagglutination inhibition (h.i.) test. When serum, containing antibodies, is mixed with a haemagglutinating virus the antibodies can prevent the haemagglutination. This phenomenon is called haemagglutination inhibition.

Twofold dilutions of pretreated serum, row 1 to 12, are made, an exact amount of virus is added and after incubation an erythrocyte suspension is added to detect haemagglutination.

Row G and H are the controls. In G no virus is added to detect non-specific agglutination of r.b.c. In H no serum is added to control the agglutination by the virus. Serum I is negative; sera 5 and 9 have a titre of 16; sera 3, 4 and 11 : 32; sera 7, 8 and 12 : 64; serum 2 : 128; serum 10 : 256 and serum 6 a titre equal to or higher than 512.

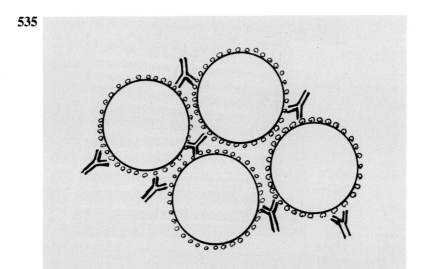

535 Passive or indirect agglutination. Inert particles like latex and tanned erythrocytes can be coated with an antigen. Addition of specific antibodies causes the coated particles to agglutinate. Pretreatment of the serum to enhance the specificity of the reaction may be necessary.

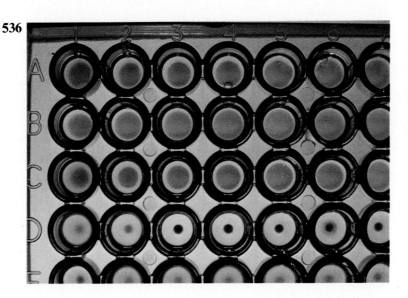

536 Latex test in a U plate. Four sera, diluted from left to right. The discolorations in some wells on the left are caused by the colour of the serum sample. The upper 3 rows are positive sera, with a pattern of sediment on the bottom. The lower row contains dilution of a negative serum; the latex particles are a sediment dot.

537 Latex particles. Coated latex particles before addition of an antiserum. The latex suspension sediments to a dot.

538 Agglutinated latex particles. Addition of an antiserum to coated latex particles causes these particles to clump and sediment in a pattern.

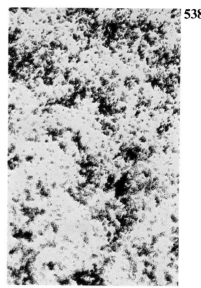

539 Heterophil antibody spot test (1). The Paul-Bunnell-Davidsohn (PBD) test, which indicates heterophil antibodies in mononucleosis, can be performed as a spot test. A drop of patient's serum, or blood, is mixed with guinea pig kidney extract to elimate the non-specific Forssman antibodies.

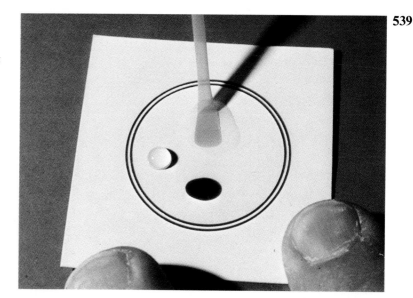

540 Heterophil antibody spot test (2). The pretreated serum is mixed with a suspension of sheep or horse erythrocytes and mixed well. In this commercial test the cells are stained blue to enhance the accuracy of the test reading.

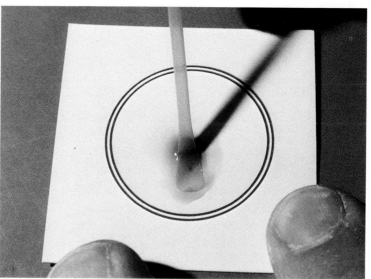

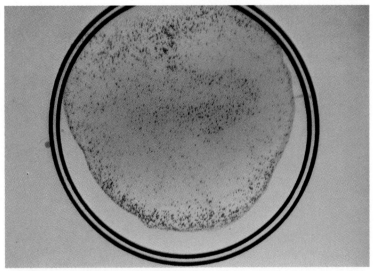

541 Positive spot test. In a few minutes an agglutination takes place when the serum is positive. Determination of the PBD titre must be done in a tube test. Specific antibody reactions to Epstein-Barr virus can be done if a specific diagnosis is needed.

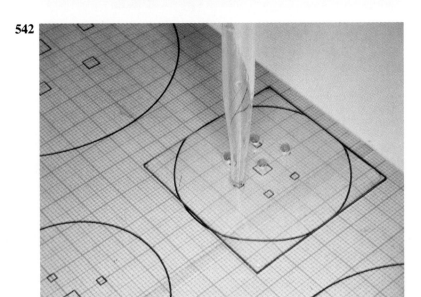

542 Immunodiffusion. Immunodiffusion in agar is used for precipitating sera and antigens. The agar gel is poured onto a pretreated slide, in this case 5 x 5 cm lanternplate coverslips. To punch the holes a pasteur pipette of the right diameter is used, guided by a drawn pattern. A template of silicone smeared plastic may be used instead of punching.

543 Performing the test. For a serum titration the antigen is added to the central hole and the serum dilutions to each of the peripheral holes. The slides are put away overnight in a humidified chamber. Two changes of saline are used to wash the slides, then they are rinsed in distilled water and dried at 37°C, covered with a sheet of filterpaper soaked in saline.

544 Staining the immunodiffusion slide. The dried slide is immersed in amidoblack for 10 minutes and then washed in several changes of 2% acetic acid in distilled water. Dry in the air.

545 Precipitation lines. Two antisera and the pre-immunization sera were tested against the antigen used for immunization. The firm precipitation line is accompanied by weaker lines due to impurities in the antigen.

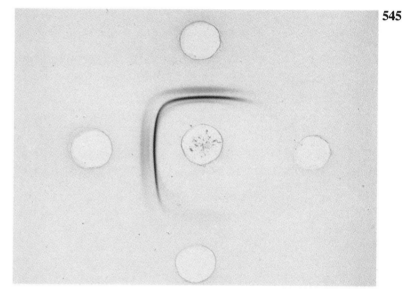

546 Haemolysis in gel (H.I.G.). Reactions. Haemolysis in gel is a suitable test for screening for the presence of antibodies. Holes are punched in an agarose plate containing complement and sheep erythrocytes coated with antigen. If the serum added to the hole contains antibodies the erythrocytes are haemolysed. The size of the haemolytic zone is related to the titre of the antibody.

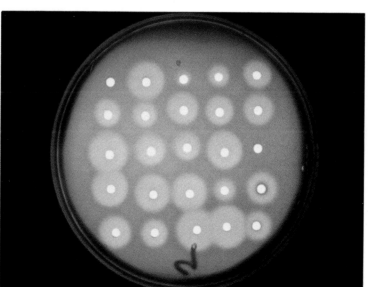

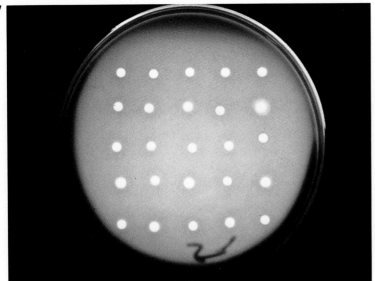

547 Haemolysis in gel, controls. All sera are also tested in a control plate, which contains all reagents except the antigen, in order to detect non-specific haemolysis which can be due to contaminants in the serum.

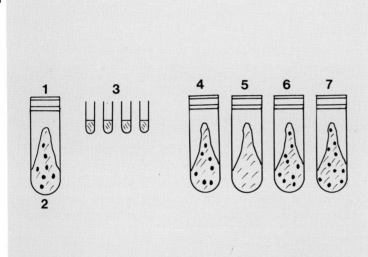

548 Identification of an isolated virus.

1 Tissue culture tube in which the virus is isolated, c.p.e. visible
2 Slide made from the tube, type of c.p.e. indicates the group to which the virus belongs
3 To type, the virus is mixed with type-specific antisera
4, 5, 6, 7 Tissue culture tubes inoculated with the serum-virus mixtures. 4, 6, 7 Tubes showing c.p.e. 5 No c.p.e. = neutralization, indicates the type of the virus

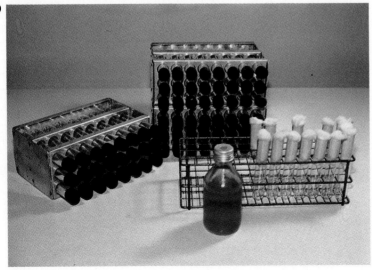

549 The neutralization test. The neutralization test is the most specific antibody test in virology to indicate the presence of immunity. The test done in t.c. tubes is labour consuming and expensive. The illustration shows the 72 tubes, and other utensils for a neutralization test on 3 sera in triplicate, to compare with the microtitre performed test, done on 4 sera.

550 Neutralization test in microtitre plates.
Three twofold dilutions are made from each serum, starting at A ending at G. H contains undiluted serum as a control. All dilutions are made in the microtitre plate. To each well, except row H, 100 TCD 50 of virus is added and the plate is incubated for one hour. Then to each well the cells are added. After 5-7 days the plate is coloured and read.

Serum number one, rows 1, 2 and 3, neutralizes three dilutions; serum number 2, rows 4, 5, 6, must be considered negative. Sera 3 and 4 show a neutralization up to 128 and 64. Exact titres must be calculated by the method of Reed and Meunch.

FLUORESCENT ANTIBODY TECHNIQUES

551 Direct immunofluorescence.

1 Cell containing...
2 Viral antigen
3 Fluorochrome labelled antibody
4 Complex of antigen and labelled antibody can be detected by...
5 Ultraviolet light

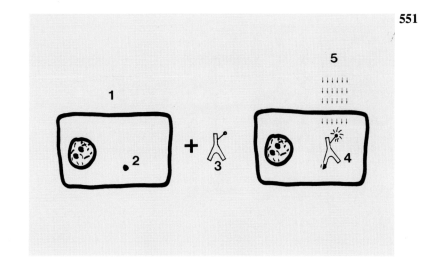

552 Indirect immunofluorescence.

1 Cell containing...
2 Viral antigen
3 Antibody against virus
4 Antigen antibody complex
5 Fluorochrome labelled antibody
6 Complex consisting of antigen, antibody, anti-antibody and fluorochrome lighting up under,
7 Ultraviolet light

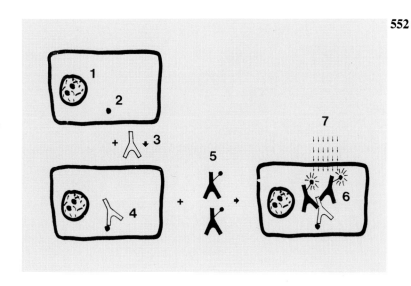

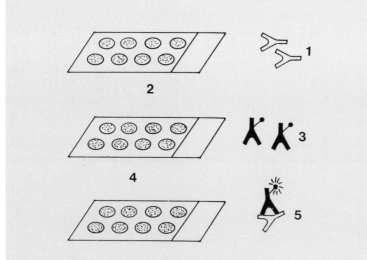

553 Fluorescent antibody technique for the detection of antiviral antibodies. Slides with wells coated with virus antigens are used.

   1 Dilution series of serum are applied to the antigen spots
   2 Wash gently after incubation
   3 Treatment with FITC conjugated antigammaglobulin (animal origin)
   4 Incubate and wash
   5 Inspect the spots with a fluorescent microscope; all spots showing fluorescence are positive for antibodies

554 Slides for the fluorescent antibody technique. Slide for the fluorescent antibody technique (FAT) are made by smearing (or growing) virus containing cells on glass. Special slides are used, printed with a pattern of wells to which the antigen is applied.

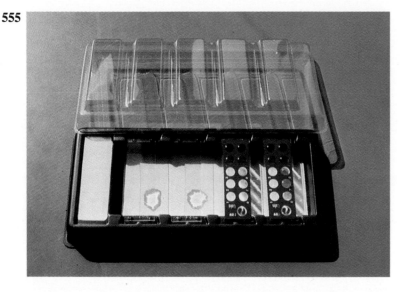

555 Incubation of slides. Slides on which a reaction takes place have to be incubated in a moist chamber in order to let the antibody bind to the antigen. Special models are available but simple petri dishes with a moist filter paper are as good.

556 Washing the slides. After each antibody reaction the slides have to be washed with several changes of saline. This is best done by placing the slides in a holder and moving them through a row of glass cuvettes with p.b.s.

After washing and drying, buffered glycerol is applied to the slide and a thin coverslip laid on top.

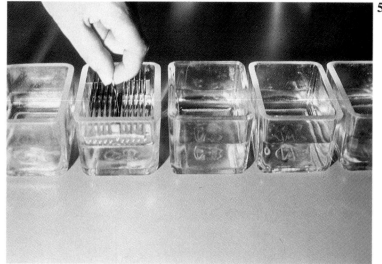

557 Fluorescence microscope. Microscopes for fluorescence work can be adapted from any ordinary model. Special models, however, are more dependable and convenient. The illumination can be done with darkfield or with incident light. The model shown illuminates the specimen from above. Two sets of filters are used, exciter filters for the illumination of the specimen and barrier filters to protect the eyes of the worker.

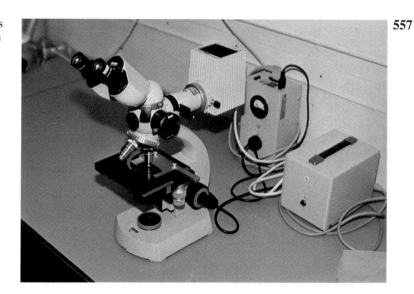

558 Antibody detection. Antibodies against Epstein-Barr v.c.a. (virus capside antigen) were assayed in this smear of cultured E.B.V. infected lymphoid cells. The positive cells show green fluorescence, indicating the presence of antibody in the serum specimen. The background is counter-stained red by Evans blue.

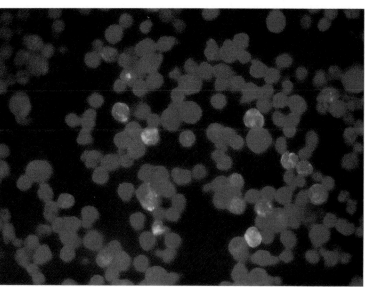

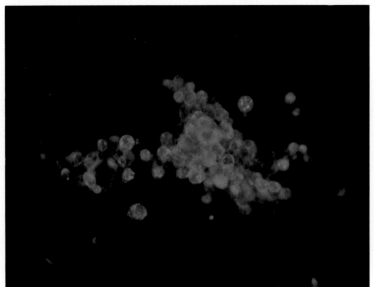

559 FAT for herpes 2 antibodies. To determine whether the patient had any antibodies against herpes type 2, a fluorescent antibody assay was performed on herpes 2 infected cells as a substrate. A positive result is shown.

ELISA TECHNIQUES

The ELISA (enzyme-linked immunosorbent assay) techniques are getting more and more important in diagnostic virology. These tests not only replace a number of cumbersome 'classical' serological techniques but have also widened the scope of the detection methods of viruses and their antigens.

The basic principles of the ELISA test are as follows. Antigens solubilized in an appropriate buffer can be coated on a plastic surface like polystyrene. When serum is added, antibodies can attach to the antigen on the solid phase. The presence of these antibodies can be demonstrated with the help of an anti-antibody, e.g. antihuman gammaglobulin. This anti-antibody is conjugated to an enzyme, like peroxidase. Adding a substrate, hydrogen peroxide and benzidine, will detect the amount of bound antibody by a degree of discoloration. This discoloration can be quantified with suitable apparatus.

ELISA reactions can also be used for the detection of antigens. In these cases the specific antibodies are attached to the solid phase and the extract containing the antigen is added. Adding an enzyme-linked antibody and a substrate leads to a colour reaction, the intensity of which is related to the amount of antigen present in the material. An example is the detection of rotaviruses in stools. Antigens can also be labelled. Enzyme labelled antigens are suitable for detecting antibodies.

Although the basic ELISA techniques are simple, the practical application can be very tricky for the inexperienced worker. For routine laboratories the standardization of the ingredients is a constant concern. This has led to a number of commercial kits for the assay of antibodies, like rubella, cytomegalo and hepatitis B, in human serum and for the assay of viral antigen in human materials.

Peroxidase techniques are available for the detection of viral antigens in cell smears and histological preparations. The antigen antibody complex in the cell is made visible by using a specific antiserum, an anti-immunoglobulin antibody conjugated with peroxidase and a substrate which produces a brown stain that will stay in situ.

THE RIA TEST

The RIA (radio immunoassay) is based on the same principles as the ELISA, except that a radio labelled antibody is used instead of an enzyme linked antibody. When the reaction has taken place the amount of radiolabel, for example ^{125}I, retained in the test is measured by a gamma counting device.

The RIA test is less and less used because of the restrictions imposed on work with radio-active substances and because in most cases an ELISA can be used instead of the RIA, giving the same information at lower costs and is less risky for the worker.

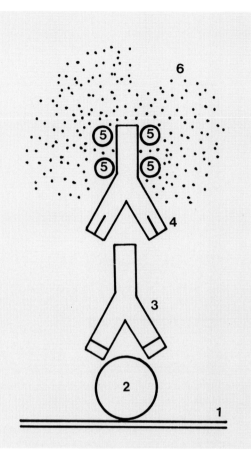

560 the ELISA reaction (enzyme linked immuno sorbent assay). Antigen is coated on a solid phase, in most cases the polystyrene surface of a flat bottomed microplate, in the Abbott system on plastic spheres. When serum is added the specific antibodies will attach to the antigen. This anchored antibody can be detected by adding an anti-antibody to which an enzyme is attached, enzyme conjugated anti human gammaglobulin. In positive cases the enzyme is still present after washing the solid phase; when a substrate is added a colour change will indicate the positive ELISA.

1 Solid phase
2 Antigen coated on the solid phase
3 Antibody captured on the antigen
4 Antibody conjugated to an enzyme and directed to the antibody captured on the antigen
5 Enzyme coupled to antibody
6 Substrate visibly changed by substrate action

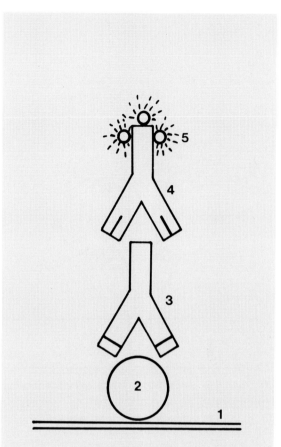

561 The RIA reaction (radio immunoassay). The principle of the RIA reaction is the same as for the ELISA except that a radio labelled antibody is used and the positive reaction is expressed by a high scintillation count. Working with radioactive substances is dangerous and to avoid risks must be done under strict controls.

1 Solid phase
2 Antigen coated on the solid phase
3 Antibody from the fluid phase attached to the antigen
4 Isotope conjugated antibody directed against the patient's antibody
5 Radio-active isotope.

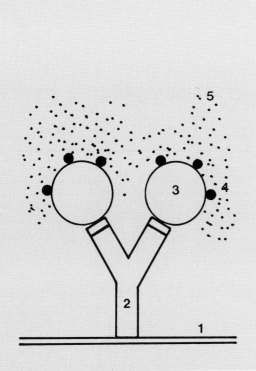

562 ELISA with antibody from serum coated to solid phase.
Theoretically it is possible to perform an ELISA by adsorbing the antibodies from the serum directly onto the solid phase and assaying for the presence of specific IgG with a labelled antigen. The disadvantage is, however, that the IgG from the serum is adsorbed non-selectively and the specific antigen to be looked for might be too small a fraction to obtain dependable results.

1 Solid phase
2 Antibody from serum coated on the solid phase
3 Antigen conjugated with enzyme
4 Enzyme
5 Substrate visibly changed by enzyme action

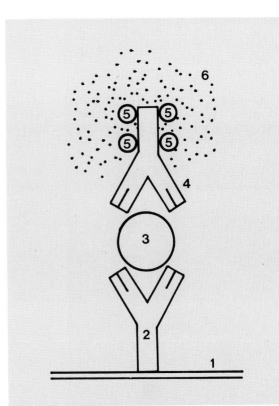

563 ELISA for detecting antigen. Coating the solid phase with a specific antibody – either monoclonal or as a high titred antiserum – enables us to capture antigens from materials like faeces, urine, blister fluids, serum, cerebrospinal fluid and so on. The captured antigen is spotted by the same antibody used for the coating but now linked to an enzyme. Practical application: rotavirus and hepatitis A virus detection in faeces

1 Solid phase
2 Antibody to the antigen searched for, coated on the solid phase
3 Antigen captured from the fluid to be investigated
4 Antibody to the antigen conjugated to an enzyme
5 Enzyme coupled to antibody
6 Substrate visibly changed by enzyme action

564 Competition method for detection of antigen by the ELISA method. Labelled antigen and extract from patient's material are added to a plate coated with antibody against the antigen. Labelled antigen causes a reaction; patient's unlabelled antigen can block the antibodies for the labelled antigen.

In this figure the patient's antigen is abesent, the labelled antigen gives the reaction; this is a negative test result. The colour reaction is strong.

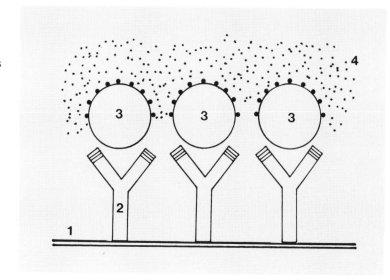

565 Positive test result. The patient's material contains antigen which attaches to the antibody, leaving little or no place for the labelled antigen. The colour reaction is weak or absent.

1 Substrate
2 Antibody
3 Labelled antigen
4 Colour reaction
5 Patient's antigen

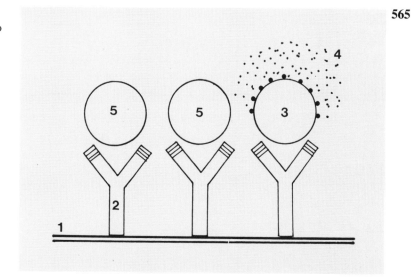

566 The Elisa test in a microplate. Microtitre plates of special quality and with flat bottomed wells are used. The colour produced is the basis for interpretation of the reaction, which is read by a photometer. The four rows on the left are the controls. The colour seen here is typical of phenylenediamine as a chromogenic substance.

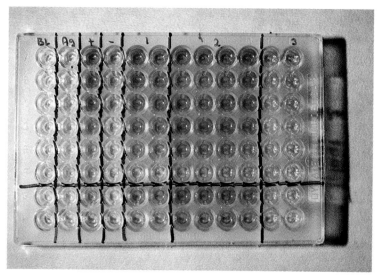

567 The ELISA test, controls.
Determination of the exact value of the reaction by the photometer is only possible when sufficient controls are built in. There must be a blank control, an antigen control, and two serum controls: a positive and a negative.

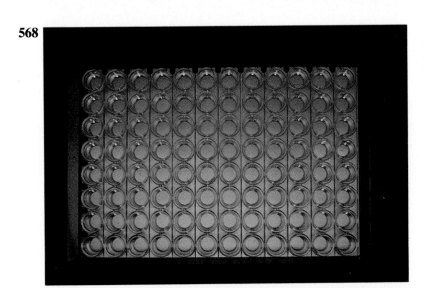

568 The ELISA test. The chromogenic substance added is dependent on the enzyme used. In these strips the colour is that of p-nitro phenylphosphate.

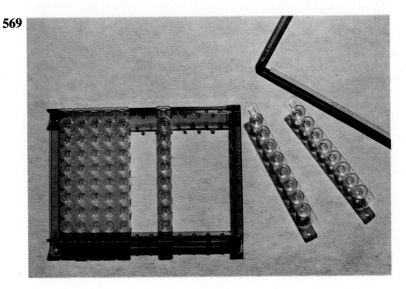

569 Flat-bottomed strips for ELISA. Using the 96 well plates is an advantage for serological reactions which must be done with large numbers of samples. When only small numbers have to be done it is more economic to use coated strips of 8 or 12 wells.

570 Washing ELISA plates. Washing is an essential step in performing an ELISA test. Apparatus such as the multiwash and microplate washer can be very helpful to speed up operations and to prevent variations due to manual operation.

571 Reading the ELISA test. Interpretation of the reaction colours of an ELISA test cannot be made by visual estimation. Special apparatus containing filters, photometers, microprocessors and so on are indispensable for useful results. A 96 well plate can be read and the results printed in 60 seconds.

572 The ELISA system, Abbott. The ELISA system devised by Abbott uses a coated plastic bead as the solid phase. This brings the advantage that a number of antigens can be used at a time and individual service is possible. The indicator system is peroxidase based and reading is done in tubes in a computerized spectro photometer.

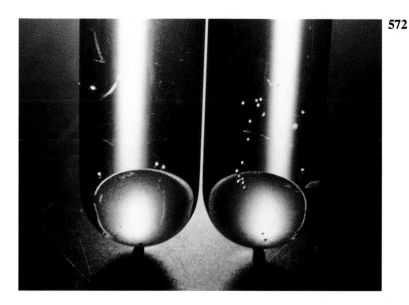

573 The ELISA reaction on beads in plates.
The beads can be used for determination of antibody titres or for the detection of antigens. In this illustration faecal extract is added to rota-antibody coated beads to diagnose a rotavirus infection.

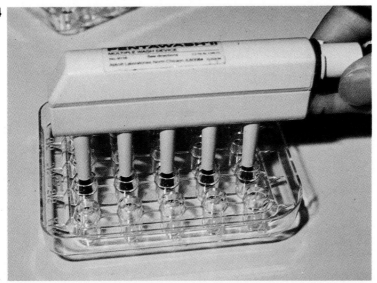

574 Washing the beads in plates.
After the reaction the beads are washed in buffer, using a device that pumps and sucks. After washing the beads are brought over in tubes.

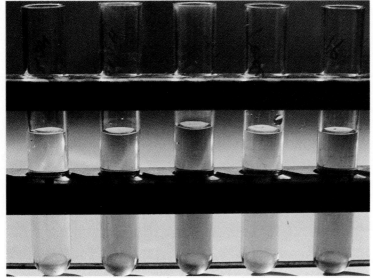

575 Enzyme-substrate reaction in tubes.
The beads that were washed in the plates are brought over in tubes and the enzyme substrate reaction is performed in the tubes. The colour is an indication of the strength of the reaction.

576 Computerized spectrophotometer. The tubes are read individually in the computerized spectrophotometer. The results being processed in the computer are based on the reading of controls which must always be included.

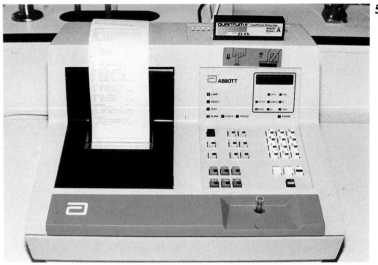

IgM TECHNIQUES

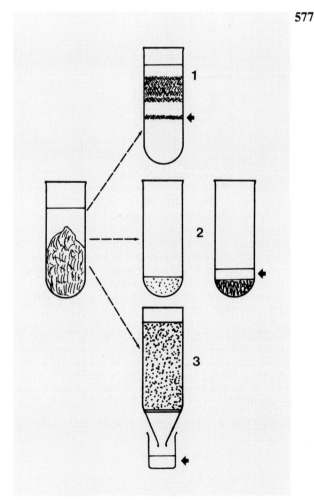

577 Separation of the IgM fraction from serum. For the assay of virus specific IgM the separation of the IgM fraction from the IgG of the serum is important. This can be performed in a number of ways.

1 Serum is put atop of sucrose gradient and is centrifuged into bands in a swing-out rotor in the ultracentrifuge
2 Serum is mixed with anti-Fc and brain egg-yolk mixture and is absorbed with liver powder
3 Serum is passed through a column and washed out, the IgM is eluted and collected. The FPLC system is a good example of this separation system.

578 IgM-IgG fractionation by ultra centrifugation, diagram.
10,20,30,40:layers of different sucrose concentrations
S:serum mixed with FITC layered on top
18 hours at 35000 rpm
IgG:layer of IgG, IgM:layer of IgM (seen in UV light)

579 Swing out rotor used for IgM-IgG fractionation by ultracentrifugation. A sucrose gradient is prepared and serum mixed with FITC is layered on top. A swing-out rotor is used to centrifuge the tubes at 35000 rpm for 18 hours. In the swing-out the tubes are horizontal during centrifugation, permitting a perfect layering that is kept intact when, after centrifugation, the tubes come vertical again.

580 Sucrose gradient tube. After centrifugation the tubes are inspected in filtered U.V. light (the worker must wear safety spectacles). The broad top band is the IgG fraction, the narrower lower band contains the IgM fraction. Perforate the tube with a needle and remove the IgM by suction with a syringe. **Caution: This IgM fraction may contain rheumatoid factor, which is IgM and may interfere with some reactions.**

581 Preparation of IgM by the anti Fc method. The IgG fraction of the serum can be removed by addition of sheep anti Fc antibodies. The precipitation being enhanced by addition of calf brain-egg yolk mixture and shown in this figure. Incubate overnight at 4°C. This method does remove rheumatoid factor.

582 Working diagram of the anti Fc method.

1 Serum + anti Fc + brain egg yolk overnight at 4°C. IgG is complexed to anti Fc.
2 Liver powder is added, absorb one hour
3 Centrifuge 30 min at 3000 rpm: IgM in supernatant, IgG-anti Fc-liver powder is sedimented.

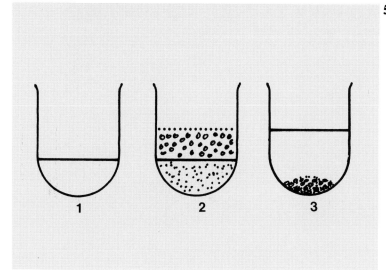

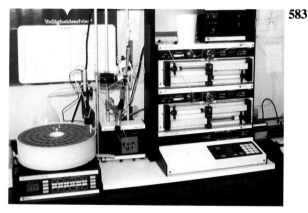

583 Column fractionation of IgM. IgM can be separated from the serum by column fractionation. Very simple devices can be used, but more consistent results are got by using a FPLC apparatus (fast protein liquid chromatography) which will fractionate the serum in a short time with great precision. The IgM fraction may contain rheumatoid factor, which is IgM, so it must be checked.

584 IgM capture technique. Anti IgM is adsorbed on the solid phase. When serum is added all kinds of IgM will be attached. To assay the presence of a specific IgM the matching antigen is enzyme labelled with the enzyme acting on the substrate.

In this way the reaction is made so specific that rheumatoid factor does not interfere as it will not bind the labelled antigen.

If an haemagglutinating antigen is used, for example measles, the reaction can be evaluated by haemadsorbtion. (SPIT – solid phase immuno sorbent technique)

1 Solid phase
2 Anti IgM antibodies adsorbed on the solid phase
3 IgM captured from the patient's serum
4 Antigen conjugated to an enzyme
5 Enzyme
6 Substrate hydrolyzed by the enzyme

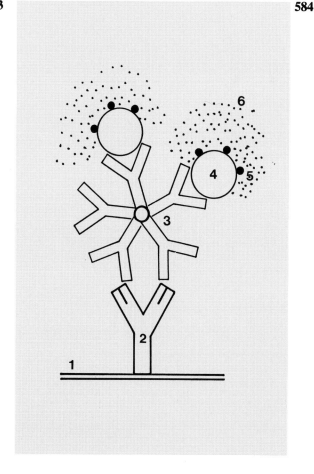

MONOCLONAL ANTIBODIES

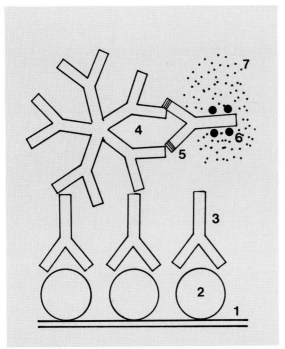

585 Interference by rheumatoid factor. When an antigen coated solid phase is used to assay IgM with an antigen IgM conjugate, false results can be obtained when rheumatoid factor is present. The reason is that IgG attaches to the antigen and the rheumatoid factor attaches to the IgG, so that the anti IgM conjugate will react with the rheumatoid factor and this will be a false positive IgM reaction.

1 Solid phase
2 Antigen adsorbed on solid phase
3 IgG from patient's serum attached to the antigen
4 IgM – rheumatoid factor attaching to the IgG
5 Antibody to IgM conjugated with an enzyme
6 Enzyme
7 Substrate hydrolyzed

586 Making monoclonal antibodies.

  1 Immunization of mice by the intraperitoneal route
  2 When serum shows sufficient antibodies, the mouse is killed and the spleen harvested
  3 Spleen is cut up and made into a cell suspension
  4 Mouse myeloma cells from a culture are added to the…
  5 Spleen cells and the cell mixture is fused by slowly adding…
  6 Polyethylene glycol (P.E.G.)
  7 The fused cells are grown in microplates or in…
  8 Soft agar. Cell colonies which produce the desired antibodies are…
  9 Selected and passaged a few times to get stable clones which are…
  10 Kept in tissue culture for production of monoclonal antibodies or are kept for use in liquid nitrogen
  11 Intraperitoneal injection of the clones induces antibody rich ascitic fluid in mice

587 Monoclonal antibodies (1). The basis for the production of monoclonal antibodies are spleen cells producing the required gammaglobulin to the antigen. Mice are immunized as a rule by the intraperitoneal route, the antigen being mixed with complete Freund adjuvant. A booster dose via the same route or by the intravenous route may be necessary.

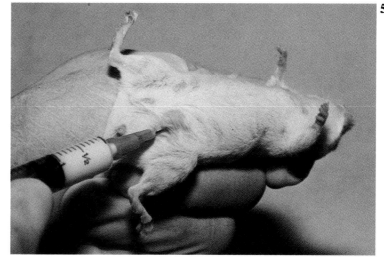

588 Monoclonal antibodies (2). The humoral immune response can be checked by taking blood samples through orbital puncture. As soon as the immunization is successful the mice are killed by cervical dislocation and the spleen is taken out.

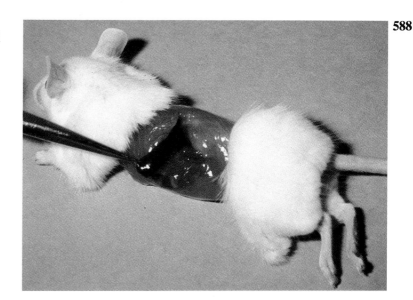

589 Monoclonal antibodies (3). The spleen is freed from adhering fat tissue and cut into small pieces. The cells are suspended and sedimented by centrifugation and washed in RPMI 1640 medium.

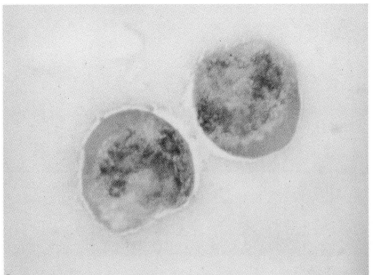

590 Monoclonal antibodies (4). Mouse myeloma cells from a line which is HAT (hypoxanthine-aminopterin-thymidine) sensitive, thus requiring exogenous hypoxanthine and thymidine. These cells are used to fuse with the antibody-forming cells, resulting in genetic complementation and an antibody-producing hybrid cell-line.

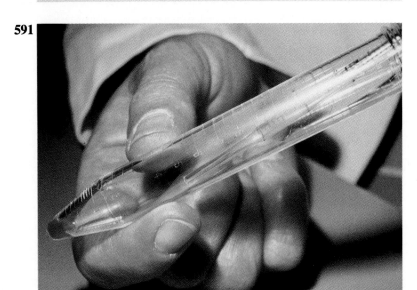

591 Monoclonal antibodies (5). The immune lymphoid cells from the spleen and the myeloma cells are mixed at a 10:1 ratio and fused by adding dilutions of PEG (polyethylene glycol) slowly while stirring. The cells are sedimented by centrifugation, mixed with HAT medium and suspended in soft agar or plated out.

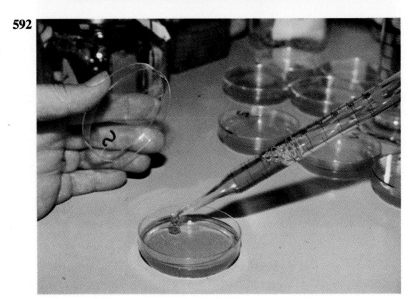

592 Monoclonal antibodies (6). Plating out of the fused cells is done by mixing them with soft agar which is poured out in petri dishes already containing a layer of feeder agar.

593 Monoclonal antibodies (7). The fused cells start growing in a few days, forming small colonies in the agar. These hybridomas produce antibodies, but only a small number are the exact type of antibody that was aimed for. The cell clones can be picked out randomly and tested for antibody production.

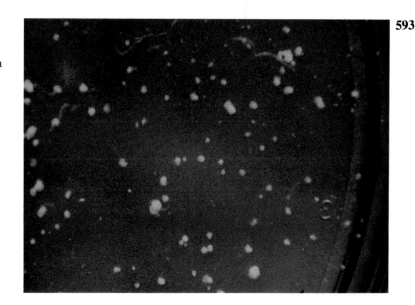

594 Monoclonal antibodies (8). Covering the soft agar layer containing the cell clones with a double nitrocellose filter – one containing the antigen, the other the control antigen – causes serological reactions to take place exactly above the producing clone. The reaction is made visible by an immunoperoxidase reaction. The exact spot is punched in the filter and the clone retrieved through this hole.

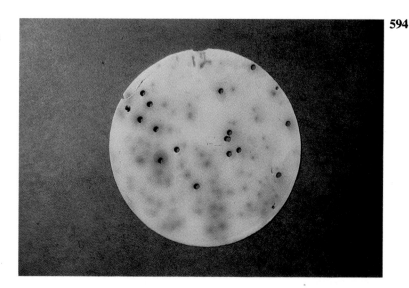

595 Monoclonal antibodies (9). The clones producing the right antibodies are further cloned during three passages in mictrotitre plates in order to get stable lines, as chromosome loss occurs with subsequent cell division.

596 Monoclonal antibodies (10). The culture fluid from the tissue culture plates with the clones are tested in the ELISA test and the best clones are picked for further work. As the hybrid strain can be lost during in vitro passage, a sufficient amount of cells must be stored in liquid nitrogen.

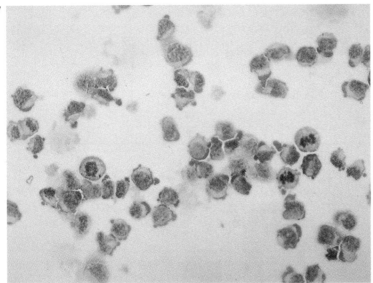

597 Monoclonal antibodies (11). Antibody production can be done in tissue culture, the supernatant can contain up to 60 microgram gammaglobulin per millilitre. Production can be expanded easily by normal tissue procedures. If necessary the antibodies can be purified and concentrated.

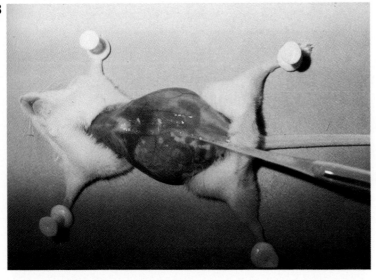

598 Monoclonal antibodies (12). A greatly increased antibody production can be attained by infecting mice with the hybridoma strain, either to induce a subcutanous tumour or to induce an ascites tumour. Levels of at least 25 milligram antibody per millitre can be reached.

STORAGE OF SERUM SAMPLES AND ERYTHROCYTES

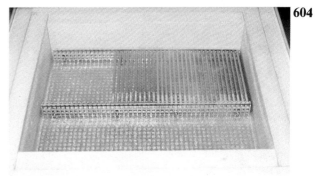

599 Serum samples. Serum samples are best stored in a standard screw cap bottle. Do not overfill. Put the identification on the bottle wall and on the top of the screwcap. This makes retrieval from the storing system easier and prevents the mixing up of samples from different patients.

600 Storing serum samples (1). In a virological lab it is important to store serum samples in a systematic way so that they can be retrieved easily. It is also important to store the samples for a long time, at least a year and if possible 5 or 6 years.

601 Storing serum samples (2). Serum samples are stored at $-20°$ to $-35°C$. Store the more recent samples in a different freezer than those which are to be stored for longer times. Do not use self de-icing machines as the temperature differences are detrimental to the quality of the serum.

602 Storing erythrocytes at 4°C. Erythrocytes needed for the c.f.t., the haemagglutination test and the haemadsorption test can be stored as blood in Alsever solution (left). A portion of the sediment is taken out and washed 3 times in saline before use. Most red blood cells keep well for several days as a 0.5 or 1.0 % suspension in saline (right). When traces of haemolysis show, discard the suspension.

603 Storing erythrocytes in liquid nitrogen (1). Storing erythrocytes in liquid nitrogen gives the advantage that for many months a constant batch can be used. The red blood cells from Alsever blood are washed and mixed with a freezing mixture containing glycerol and sorbitol. The suspension is sucked into plastic tubes (pailets).

604 Storing erythrocytes in liquid nitrogen (2). The tubes, used in a different colour for each kind of erythrocytes, are immersed in liquid nitrogen, snap freezing the contents. The isolating tray shown was custom-made.

605 Storing erythrocytes in liquid nitrogen (3). The snap frozen tubes containing the erythrocytes are stored in liquid nitrogen in double walled vacuum vessels. Handle carefully, prevent frost bite and wear safety spectacles.

606 Thawing frozen erythrocytes. To make a suspension of the erythrocytes stored frozen in pailets, thawing must be quick. Put as many tubes as needed in a calibrated tube immersed in tepid water. Wash the erythrocytes to remove the haemolysed red blood cells.

LYOPHILIZATION

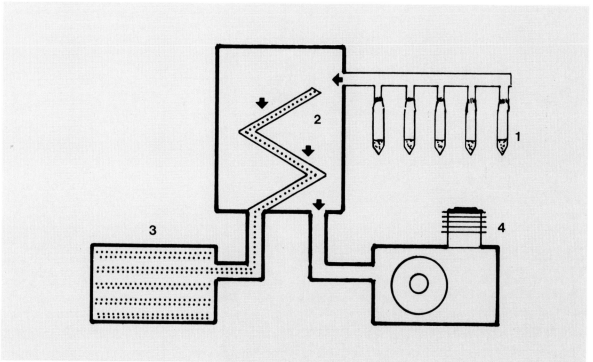

607 The principles of freeze-drying (lyophilization). Water evaporates quickly when frozen and under vacuum, the evaporation being so intense that the fluid is kept in a frozen state without additional cooling. This principle is used to freeze-dry organic materials, for example serum, antigens, viruses, trypsin and antibiotics.

The materials to be dried are put into ampoules, the contents are frozen and the ampoules coupled to the lyophilization apparatus. Now the ampoules are drawn vacuum which starts the evaporation of the water. The water vapour is trapped in a moisture trap with a cooling spiral to maintain an optimal vacuum and to keep the oil of the pump dry.

As soon as the materials are dried the top of the ampoules are sealed by melting the glass. Test the ampoule for vacuum on the day of sealing and a week later. Leaky ampoules have spoiled contents.

After freeze-drying contagious materials the apparatus, including the moisture trap, must be disinfected as small dried particles can be dispersed into the circuit.

1 Ampoules to be freeze-dried
2 Moisture trap with refrigerated cooling spiral
3 Refrigeration machine
4 Vacuum pump

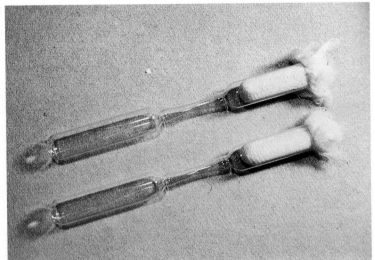

608 Ampoules for lyophilization. Many materials, like serum, organ suspensions, enzymes and so on, can be dried from the frozen state. Ampoules for this purpose come in many models. The type shown here can be sterilized by plugging with cotton and heating in an oven. The amount to be processed is up to 2 millilitre.

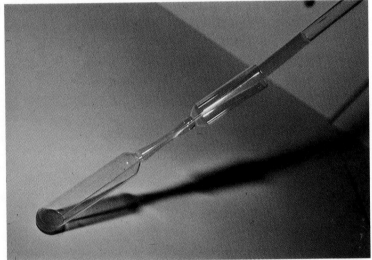

609 Filling an ampoule. Ampoules are filled by carefully depositing the fluid at the bottom. Avoid contamination of the narrow part as this part is to be heated for closing the ampoule and contamination can interfere with the vacuum.

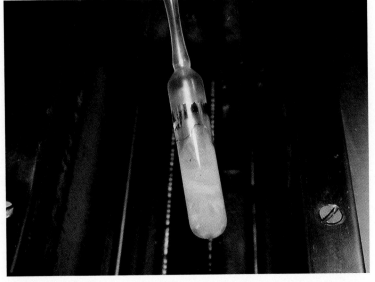

610 Freezing the ampoule. The ampoule is frozen in an alcohol bath of minus 25-30°C. If the contents are more than 0.5 ml, it is better to rotate the ampoule during freezing in such a way that the greater part of the fluid freezes to the wall. This enlarges the surface of the dried product, enhancing reconstitution.

611 Drying from the frozen state. The ampoules with the frozen contents are coupled to the freeze-drying machine. The vacuum in the ampoule causes the water to evaporate and this lowers the temperature to such a degree that the contents remain frozen.

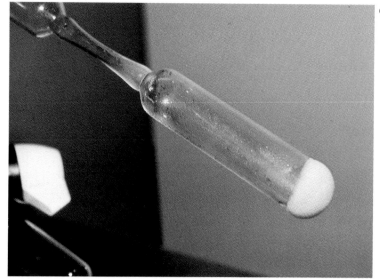

612 Freeze-dried ampoules. Ampoules can be dried individually or on adapters which allow drying of 12 or more ampoules. In 8 to 10 hours, depending on the capacity of the machine, the contents of the ampoules are dry.

613 Sealing under vacuum. As soon as the contents are absolutely dry the ampoules can be sealed. A high intensity gas-burner is used to heat the narrow part of the ampoule.

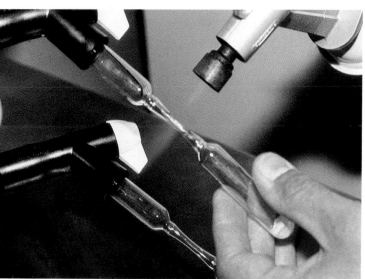

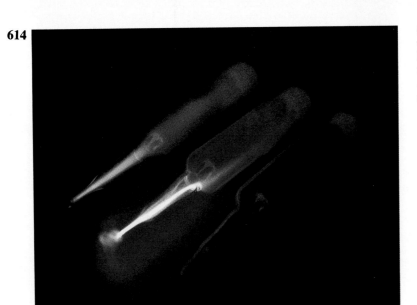

614 Testing the vacuum. To test the vacuum a high frequency tester is used. The h.f. current causes the ampoules, in which the vacuum is satisfactory, to light up. Ampoules that are insufficiently sealed leak air and will not light up.

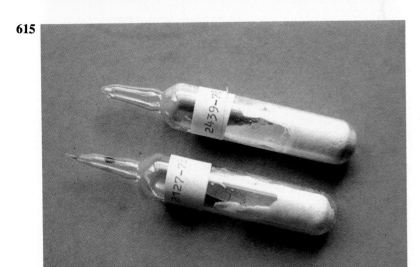

615 Freeze-dried serum. These ampoules show the appearance of freeze-dried serum. The serum is spread over the wall and is finely porus. A common mistake made is not freezing the ampoules on the machine. In that case the serum starts 'boiling', contaminating the machine and resulting in an unreliable product.

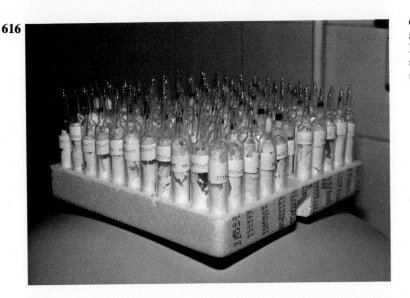

616 Storing ampoules. Freeze-dried ampoules of serum can be stored at 4°C. Pack them so that the sealed tips are not severed, otherwise air will come in and spoil the contents.

Further reading

Belshe, R.B. (Ed),
Textbook of Human Virology,
P.S.G. Publ. Co. Littleton, 1984.

Freshney, R.I.,
Culture of Animal Cells,
Alan R. Liss., Inc. New York, 1983.

Howard, C.R. (Ed),
New Developments in Practical Virology,
Alan R. Liss., Inc. New York, 1982.

Kawamura, A. and Aoyama, Y. (Eds),
Immunofluorescence in Medical Science,
Springer Verlag Berlin New York, 1983.

Lenette, E.H. (Ed),
Laboratory Diagnosis of Viral Infections,
Marcel Dekker Inc. New York, 1985.

Lenette, E.H. and Schmidt, N.J.,
Diagnostic Procedures for Viral, Rickettsial and Chlamydial Infections, (5th. ed.),
American Public Health Association. Washington, 1979.

Malherbe, H.H. and Strickland-Cholmley, M.,
Viral Cytopathology,
C.R.C. Press. Boca Raton, 1980.

de la Maza, L.M. and Peterson, E.M. (Eds),
Medical Virology I, II and III,
Elsevier, Amsterdam, 1982, 1983, 1984.

McLean, D.M.,
Immunological Investigation of Human Virus Diseases,
Churchill Livingstone, London, 1982.

McLean, D.M. and Wong, K.K.,
Same-day Diagnosis of Human Virus Infections,
C.R.C. Press. Boca Raton, 1984.

Palmer, E.L. and Martin, M.L.,
An Atlas of Mammalian Viruses,
C.R.C. Press. Boca Raton, 1982.

Edmond, R.T.D.
A Colour Atlas of Infectious Diseases,
Wolfe Medical Publications, London, 1984.

Index

All figures refer to page numbers

A

Acetylamine fluoride (AAF) 186
Adenovirus 180
   intranuclear inclusion body 146, 147
   particles 173, 174
Aerosols 16, 27, 32, 39
Agglutination, indirect 204
Air sac 44, 47
Air space, artificial 55, 56
Allantoic cavity 44
   inoculating 45–49
      air sac 47
      death of embryo 48, 60
      sealing mixture 46
Allantoic fluid, harvesting 45–49
Alsevers solution 97
Aluminium foil caps, in dry sterilization 35
Amniotic cavity 44
   inoculation of 50–53
      sealing egg 51
Amniotic fluid
   harvesting 52, 53
   rubella virus in 182
Amphotericin B 99
Ampoules, for lyophilization 230, 231
Anaphylactic shock 68
Animals, experimental, use of 67–97
   collecting blood for erythrocytes 95–97
   handling laboratory animals 84–95
      ferrets 94
      guinea pigs 89
      mice 84–88
      monkeys 95
      rabbits 94
   practical uses for antisera 69
   production of antisera 68
   suckling mice 70–83

Antibiotics 99
Antibody
   fluorescent techniques 15, 209–211
   heterophil spot test 205, 206
   monoclonal 68, 179, 222-226
      and agar 224
      with Freund adjuvant 223
      HAT-sensitive 224
      and hybridoma strain 226
      and immunoperoxidase reaction 225
      and PEG 224
      and RPMI 1640 medium 223
   neutralization test 208
   peroxidase conjugated second 184
   reaction with antigen 188
   specific 68
Antigens 38, 156
   complement-fixing 188
   ELISA for 214
   relation with antibodies 188
   titration of 199, 201
Antisera
   conjugated 179
   monoclonal 188
   practical uses for 69
      control sera 69
      identification of viruses 69
      non-specific titres 69
   production of 68
      immunization procedures 68
      lyophilization 68
      specific antibodies 68
Aujeski virus 39
Autoclave 30
   control of 31
Arboviruses 67, 96
Autoradiogram 186, 187

B

Bacterial and cell contamination 138–139
Balloons, pipette 22
Blood
   collecting for erythrocytes 95–97
   sheep 97
   sample 13

Bornholm's disease, Coxsackie B virus 76
Brown adipose tissue 78

C

Cabinets
   laminar flow 18
   safety 18, 19
Candida contamination 138
Candling
   egg 38, 42–44
   lamp 41
Capture technique 69
Carbon dioxide incubator 115, 127
Cell contamination 138–139
Cell counting 112
Cell culture techniques 98–132
   aspects of normal 135–137
   cell contamination 99
   chick embryo fibroblasts 107–108
   chlamydiae in 131–132
   immunofluorescence slides 130
   methods 114–123
      collodium 120–123
   microbiological contamination 99
   microtitre plate 126–130
   monolayers 99
   mouse embryo 104–106
   perfusion trypsinization 109–113
   primary kidney 101–103
   on slides and petri dishes 123–126
   systems 100
   tissue culture media 98–132
Cell lines
   continuous 100
   diploid 100
Central European Encephalitis (CEE) virus 39, 61, 87
Centrifuges 27
   tubes 27, 36
Chamber
   Sayk sedimentation 14, 15, 133
   slide 123, 124
Chick embryo
   fibroblasts 107, 108, 135
   techniques 38–66, 44
      embryonated egg 38

harvesting allantoic fluid 45–49
inoculation of allantoic cavity 45–49
inoculation of amniotic cavity 50–53
inoculation of chorioallantoic cavity 54–66
Chickens
erythrocytes 202
heart puncture in 95, 96
Chlamydiae 39
darkfield 132
elementary and reticulate bodies 178
in cell culture 131, 132
ribosomes 178
staining, Giemsa 131, 132
trachomatis, in non-gonococcal urethritis 178
Chorioallantoic membrane, inoculation of 54–66
absence of lesions 61
artificial air space 55, 56
aspect at 17 days 59
CEE virus 61, 62
cowpox lesion 65
smear of 66
cutting lesions 59
cutting the shell 55
harvesting 58–59
herpesvirus 62, 63
lesions 63, 64
inspection of 59
louping-ill virus 62
lymphocytic choriomeningitis lesions 64
marking the eggs 57
marking the shell 55
non-specific lesions 61
normal histology of 60
vaccinia lesions 65
washing 58
Collagen 99
Collodium method 120–123
eosin 121
haematoxylin 120
Complement fixation 97
tests 189, 196–197
screening 197, 198
titration 199
Contact inhibition 99
Contaminated materials, sterilization 29, 30
Contamination, tissue culture
bacterial 138–139
candida 138
cell 139
mycoplasma 139

yeast infection 138
Cotton wool plugs, in dry sterilization 35, 36
Coxsackie A and B virus 67, 71, 73, 75, 76, 129
Bornholm's disease in 76
cytopathic effect 143
fat necrosis in 77
macroscopic appearance of 77
myositis in 76
pancreatitis in 78
Cowdry type inclusions
herpesvirus 164, 176
Cowpox 39
inclusion bodies 150, 151
lesions 65, 175
smear of 66
Cytocentrifuge 15
slide 15
Cytodex beads 117, 118
Cytomegalovirus 11, 98, 158, 161, 165, 180, 186
congenital 165
cytopathic effect of 150
immunosuppression in 166
nuclear changes in 150
Cytopathic effects 133–157
aspects of normal cell cultures 135–137
bacterial and cell contamination 138–139
non-specific 140–141
viral 142–157

D
Dapi stain 186
Darkfield examination, in chlamydiae 132
Depilatory, for collecting sheep's blood 97
Diluter
hand-operated 192
mechanically operated 193
micro 190
tips 192
Dimethyl sulphoxide (DMSO) 113
Diploid cell strains 98, 100
Disinfecting
fluid, pipettes, 24, 25
hands 28
Disposable units, for filtration 26
DNA
complementary strands 184
radio-isotope-labelled probe 184
recombinant techniques 184, 186
restriction endonuclease analysis of 184–187

–RNA hybrid 184
spot hybridization 187

E
Eagle M.E.M. 98
ECHO virus, cytopathic effect 143, 179
meningitis 181
Ectromelia 39
Egg
candling 38, 42–44
embryonated 38
harvesting 46
incubated, scheme 44
incubator 40
inoculation of 38
marking 57
sealing mixture 46, 51, 57
shell, drilling 45, 51
quality of 38
viruses grown in 39
Egg, inoculation route
allantoic cavity 45–49
amniotic cavity 50–52
chorioallantoic membrane 54–66
yolk sac 52, 53
Electron microscopy 171
grids for 172
immune (I.E.M.) 171
ELISA 171, 190, 213, 216, 217
competition method 215
for detecting antigen 214
reaction 213
on beads 218
system 217
techniques 212
Embryo
cells, mice 104–106
chick, techniques 38–66
candling the egg 42–44
fibroblasts 107, 108, 134
inoculating allantoic cavity and harvesting fluid 45–49
death of embryo 48, 60
inoculation of amniotic cavity 50–53
inoculation of chorioallantoic membrane 54–66
Embryonated egg 38
Encephalitis 87
Central European virus (CEE) 39, 61, 87
herpes 164
Japanese 39
St Louis 39
tickborne 163
Encephalomyocarditis, cytopathic effect of 145

Endonuclease, restriction analysis
    185–187
  of DNA 187
Enzyme-linked immunosorbent
    assay (*see* ELISA)
Eosin 121, 133
Eppendorf pipette 23
Epstein-Barr (EB) virus 156, 206
  antigens 156, 211
Erythrocytes
  collecting blood for 95–98
    heart puncture in chickens
        95, 96
    venepuncture in goose 96
    sheep 97
  chicken 202
  human 188
  storage of 227–228
    in liquid nitrogen 227, 228
    frozen 228
Evans blue stain 211
Extracellular matrix 99

F
Faecal suspension 12, 13
Fat necrosis, in Coxsackie B virus
    77
Ferret handling 94
  intranasal infection of 94
  orbital puncture in 94
Fetal calf serum 98, 99
Fibroblasts
  chick embryo 135
  human embryo 135
Fibronectin 99
Filtration, in virus diagnosis 26–28
  of small amounts 26
  of large amounts 26
  to separate viruses and proteins
      26
  disposable units for 26
  centrifuges 27
  aerosols 27
Flasks
  plastic disposable 115
  polystyrene 116
  Roux 115
  trypsinizing 109
Fluorescent assay test (FAT) 188,
    210, 212
Foamy virus, cytopathic effect of
    152
Foot and mouth disease 67
Freeze-drying (lyophilization)
    229–232
  ampoules for 230, 231
  principles of 229
  sealing under vacuum 231
  storing 232

vacuum testing 232
Freund's complete adjuvant 68, 223

G
Gammaray sterilization 37
Gas sterilization 23, 37
Gentamycin 99
Giemsa, staining in chlamydiae
    131, 132
Glassware, washing 31, 32
Glycerol, buffered 211
Gly-medium 9, 10
Goose, venepuncture in 96
Guinea pig, handling 89
  bleeding 89
  heart puncture 89
  orbital puncture 89

H
Haemadsorption 133, 156, 157
Haemagglutination assay, immune
    adherence (IAHA) 199, 200, 202
  inhibition test 204
  of red blood cells 201
  test 203
Haemagglutination pattern 49, 190,
    203
Haemagglutinins 38
  test 39
Haematoxylin 120, 133
Haemacytometer 112
Haemolysis in gel (HIG) 207, 208
Harvesting, allantoic fluid 45–49
Heart puncture
  in chickens 95, 96
  in guinea pigs 89
Hepafilters 18, 19
Hepatitis 159, 160, 214
  liver 159
  yellow fever 159, 160
Hepatomegaly 161
Herpesvirus 11, 39, 62, 63, 64, 68,
    71, 88, 93, 162, 164
  Cowdry type A 176
  cytopathic effect of 149
    equine 155
  in ganglion gasseri cell 165
  keratitis 93, 94
  lesions 63, 64
  particles 176
  simplex 68, 158, 161, 183
Heterophil antibody spot test 205,
    206
High pressure steam sterilization 30
Histopathology, of viral infections,
    see Viral infection,
        histopathology of
Homogenizer, grinding tissues
    in 17

Hybrization techniques 184
  in situ 184, 187
  DNA-RNA 185
  spot 186, 187
Hyperkeratosis 158

I
IgM
  class antibodies 188
  techniques 219–221
    anti-Fc method 220, 221
    capture technique 221
    column fractionation 221
    FPLC apparatus 221
    SPIT 221
    ultracentrifugation 220
Immune adherence
    haemagglutination assay 200,
    201
Immunization procedures 68
Immunocompromised patients 158
Immunodiffusion 206, 207
Immunofluoresence
  direct 209
  incubation of slides 210
  indirect 209
  microscopy 179–184
  slide cultures for 130
  in chlamydiae 132
Immunoperoxidase staining 184
Impression smear 11
Inclusion bodies 140
  in cowpox virus 150
  ademo 146, 147
  cytomegalo 150, 161, 165, 166
  herpes 149, 162, 164, 176
  measles 153
  molluscum 170
  reo 146
  RS 155
  SV40 152
  varicella zoster 149
Incubation, of I.F. slides 210
Incubator
  CO_2 115, 117
  egg 40
Infection intranasal,
  of ferret 94
  mouse 86
Infectious bovine rhino-tracheitis
    virus 155
Influenza 39, 68, 147, 157, 167
  primary 168
Inhalation narcosis, in mice 80
Injection
  of mice 85
    intramuscular, intracerebral,
    intravenous 85
    intraperitoneal, 86

of rabbit 92
    intravenous, intracutaneous 92
Inoculation
    of allantoic cavity 45–49
    of amniotic cavity 50–53
    of chorioallantoic membrane 54–66
    of egg 38
    of suckling mice 75–78
    of yolk sac 52, 53
Interscapular fat, in mice 77, 78
Isolator 86

J
Japanese B virus 39, 87

K
Keratitis, herpes in rabbits 93, 94
Kidney
    mouse cells, monolayer 134
    primary cell cultures 101–103

L
Labatory animals, handling 84–95
    ferrets 94
    guinea pigs 89
    mice 84–88
    monkeys 95
    rabbits 90–94
Labware, washing 31
Laminar flow cabinets 18–20
Latex test 204
    particles 205
Leighton tubes 119, 133
    coverslip culture 119
Liver
    hepatitis 159
    hepatomegaly 161
    herpes simplex infection of 161
    mononucleosis 161
Louping-ill virus 62
Lymphocytes, atypical 158
Lymphocytic choriomeningitis
    lesions 64
Lyophilization 68, 229–232
    principles of 229
    ampoules for 230, 231
    ampoules, storing of 232
    sealing under vacuum 231
    vacuum testing 232

M
Measles virus
    cytopathic effect of 153
    meningitis 182
Medium 98, 199
Meningitis
    ECHO virus 181
    measles 182

Mice, suckling, in diagnostic virology 70–83
    bleeding 84
    breeding 70
    Coxsackie A and B virus in 75, 76, 77
        fat necrosis in 77
    dissecting 73, 74, 81, 82
        storing samples 74
    ear-marking 72, 79–83
    embryo cells 104–106
    encephalitis in 87
    inhalation narcosis 80
    injection of 85–86
    inoculating 72
    kidney, cell culture 134
    killing 80
    paralysis in 87
    perfusion of 82, 83
    signs of illness in 75
Microbiological contamination, in cell culture techniques 99
Microbiological safety cabinet 19, 20
Microdiluter 193
Microscopy
    electron 171–178
    fluorescence 9, 209–211
    immunofluorescence 179–184
Microtitre cultures, cytopathic effect in 128
Microtitre cups 189
Microtitre plate culture 126, 130, 189
    antiglare shield 200
    centrifugation of 195
    cytopathic effect 128
    sealing tape 194
    shaker 194
Microtome, refrigerated 16
Microtray cultures 127
Mitosis
    normal 136
    abnormal 137
Molluscum contagiosum 158, 171
Monkeys 95
    cercopithecus aethiops 98
    rhesus 67
Monoclonal antibodies 222–226
    cells in agar 224
    Freund adjuvant 223
    HAT-sensitive myeloma cells 224
    hybridoma strain 226
    immunoperoxidase reaction 225
    PEG cell fusion with 224
    RPMI 1640 medium 223
Monolayers 99

contact inhibition 99
fibronection 99
of mouse kidney cells 134
non-confluent 136
Mononucleosis 158, 161
    blood smear 11
Morosow stain 10, 65
    elementary bodies in slide 66
Multiwell plate 126
Mumps 39
Mycoplasma 139
Myositis, in Coxsackie A virus 76

N
Negri bodies 158, 163
Neuronophagia 158
Neutralization test 209, 210
Newcastle disease 39
Nitrogen, liquid, storage 100
    freezing cells in 113
    of erythrocytes 227, 228
Norwalk virus 174
Nystatin 99

O
Orbital puncture
    in guinea pig 89
    in ferret
Orf (sheep pox) lesion 10
    cytopathic effect of 145
Orthomyxovirus 157
Owl's eye cells 161, 165

P
Pancreatitis, in Coxsackie B virus 78
Paper towels, dispensers 29
Papilloma virus 158, 171
Papovirus BK 185
Paralysis, in animals 87
Paramyxovirus 157
Pasteur pipette 24
    standardizing 24
    use of 24
Paul-Bunnell test 97
    –Davidsohn (PBD) test 205
Perfusion trypsinization 109–113
Perivascular cuffing 158
Perivascular infiltrates 158
Peroxidase conjugated second antibody 184
Phenylenediamine, as a chromogenic substance 215
Phosphotungstic acid 171, 176
    negative staining 172
Pipette, use of in virus diagnosis 22–25
    balloons 22
    disposable tip 191

dry heat sterilization 23
emptying 25
Eppendorf 23
gas sterilization 23
glass sterilizing 23
machine 191, 192
multiple channel 191
Pasteur 24
   standardizing 24
   use of 24
   pumps 22, 23
Plaque counting 125
   purification and reduction 125
Pneumocystis carinii 166
Pneumonia 158
   giant cell 167
   primary influenza 168
   staphylococcal, influenza 168
   varicella 169
Poliomyelitis virus 67, 143, 162, 175, 179
   vascular cuffing in 164
Polylysine 99
Poxviruses 10
Precipitation lines 207
Proteins, non-virion 68
Pumps, pipette 22, 23
Purkinje cells 163

R
Rabbit, handling 90–94
   box 91
   ear-marking 90, 91
   herpes keratitis in 93, 94
   immunization by scarification 92
   intravenous, intracutaneous injections 92
Rabies 39, 67, 163
Radioimmunoassay (RIA) 171
   reaction 213
   test 212–219
Refrigerated microtome 16
Reovirus
   cytopathic effect of 146
   inclusion body 183
   in nude mouse 88
Rhesus monkeys 67
Rheumatoid factor 189, 221
Rhodamine conjugated gammaglobulin 183
RNA
   –DNA hybrid 184
   polymerase 184
Roller bottles 116, 117
Roller drum, in cell culturing methods 114
Rotavirus particles 174, 214
Roux flask, in cell culturing methods 115

RS (respiratory syncytial) virus, cytopathic effect of 154
   inclusion bodies 155
Rubella virus 68, 96, 183
   in amniotic fluid 182
   in embryo 182

S
Safety cabinets 18, 38
   microbiological class I 19
   microbiological class II and III 20
Sayk sedimentation chamber 14, 15, 133
Screening, complement fixation 197, 198
Semliki Forest virus, cytopathic effect of 144
Sendai virus 39, 157
Serological methods 9
   safety rules in 9
Serological tests 188–232
   ELISA and RIA tests 212–219
   fluorescent antibody techniques 209–212
   IgM techniques 219–221
   lyophilization 229–232
   monoclonal antibodies 222–226
   pretreatment of sera 188, 189, 203
   storage of samples and erythrocytes 227–228
Serum
   antiserum 68, 69
   absorbing with kaolin 203
   blood samples 13
   control 69
   cytotoxic 68
   fetal calf 68, 98, 99
   freeze-dried 232
   lyophilization 229
   quality of 14
   storing samples 227–232
Sheep
   blood, collecting 97
   pox (orf) 10, 45
Sirc cells 135
Slide
   cells on 123–126
   chamber 123, 124
   cytocentrifuge 15
   cultures for immunofluorescence 130
   incubation of 210
   preparation 10, 12, 15
Sonification, of cells 18
Spectrophotometer, computerized 219
Spinner cultures 99, 117

SPIT (solid phase immunosorbent technique) 221
Stains
   dapi 186
   eosin 121, 133
   Evans blue 211
   Giemsa, in chlamydiae 131, 132
   haematoxylin 120, 133
   immunoperoxidase 184
   Morosow 10, 65, 66
Staphylococcal infection 168
Steam sterilization, high pressure 30
Sterilization 28–37
   autoclave 30,
   collecting contaminated materials 29
   control of 31, 33, 34
   disinfecting hands 28
   dry 33–36
      aluminium foil caps 35
      centrifuge tubes 36
      cotton wool plugs 35, 36
      packing for 34, 35, 36
      wrapping for 34, 36
   dry heat 23
   gammaray 37
   gas 23, 37
   insufficient 31
   steam 30
St Louis encephalitis 39
Stool, sample 12
Subacute sclerosing panencephalitis (SSPE) virus 154
Sucrose density gradient 189, 220
SV40 virus 88, 177, 181
   cytopathic effect of 152
Swabs, for virological diagnosis 9
Syncytia 133
Syncytial virus, cytopathic effect of 153
   respiratory (RS) 154, 166

T
Tickborne encepahlitis 163
Tissue culture
   cell culture systems 100–132
   contamination 138, 139
   hybridoma techniques 222–226
   non-specific cytopathic effect 140-141
   normal cell cultures, aspect 134–137
   safety cabinets 18–20
   viral cytopathic effect 142–157
   grinding 16, 17
Toxoplasma gondii 156

Tracheal epithelium 167
Traumatic ulcers on C.A.M. 39
Trypsin 99
Trypsinization, perfusion 109–113
  cell counting 113
Trypsinizing flasks 103
Typing, of virus 21

U

Ultracentrifugation 220
Ultracentrifuge 28
Ultrasonic cleaning 32
Ultrasonic disruption of cells 18
Uranyl acetate 171
Urethritis, non-gonococcal 178
Urine, sample 13

V

Vaccines 38
Vaccinia lesions 39, 65, 122, 177, 180
  cytopathic effect of 151
Vacuum
  sealing 231
  testing 232
Vancomycin 99
Varicella zostervirus 98, 158, 181
  cytopathic effect 149
  haemorrhagic 170
  lesion 169
  pneumonia 158, 169
  vesicle 169
Variola 39
Venepuncture, in goose 96
  in sheep 97
Verocells 118
Vesicular stomatitis virus 145
Viral cytopathic effect 142–157
  adeno 146, 147
  Coxsackie B virus 142, 143
  cowpox 150, 151
  cytomegalovirus 150
  diffuse type 142
  ECHO virus 143
  encephalomyocarditis 145
  Epstein-Barr 156
  foamy 152
  focal 148
  herpesvirus 149, 155
  infectious bovine rhino-tracheitis 155
  influenza 147
  measles 153
  orf 145
  of poliovirus 143

reo 146
Semliki forest 144
SV40 151, 152
syncytial 153, 154
SSPE 154
Toxoplasma gondii 156
varicella 140
vesicular stomatitis 145
West Nile 144
Viral infections, histopathology of 158–178
  electron microscopy 171–178
  hybridization techniques 184
  immunofluorescence microscopy 179–184
  restriction endonuclease analysis of viral DNA 184–187
Virology, diagnostic
  collecting specimen 9–14
  direct methods 10, 11, 179, 184, 209, 214, 215
  hybridization techniques 184–187
  serological tests 188–207
  virus isolation in:
    eggs 38–66
    animals 67–88
    cell culture 98–133
Virus
  adeno 146, 147, 173, 174, 180
  arbovirus 67, 96
  Aujeski 39
  Central European Encephalitis (CEE) 39, 61, 87
  chick embryo techniques for specific 39
  chlamydiae 39, 131–132, 178
  Cocksackie A and B 67, 71, 73, 75, 76, 129, 143
  cowpox 39
  cytomegalovirus 11, 98, 158, 161, 165, 180, 186
  diagnosis of 9–37
    filtration 26–28
    rapid methods, direct and indirect 9
    serological methods 9
  electromelia 39
  Epstein-Barr 156, 206, 211
  foamy 152
  herpes 11, 39, 62, 63, 64, 68, 71, 88, 93, 162, 164
  identification of 69
  influenza 39, 68, 147, 157, 167
  inhibitors 98

  isolation 171, 208
  Japanese B 39, 87
  latent 67
  louping-ill 62
  measles 153, 182
  molloscum 170
  mumps 39
  Newcastle disease 39
  Norwalk 174
  orf 10, 145
  orthomyxovirus 157
  papilloma 158, 171
  papovirus BK 185
  paramyxovirus 157
  poliomyelitis 67, 143, 162, 175, 179
  rabies 39, 67
  reo 88, 146, 183
  rubella 68, 96, 182, 183
  Semliki Forest 144
  Sendai 39, 157
  SSPE 154
  St Louis encephalitis 39
  SV40 88, 177, 181
  syncytial 153, 154, 166
  titration of 129
  transport medium 9
  typing of 21, 129
  vaccinia 39, 65, 122, 151, 177, 180
  varicella zoster 98, 149, 158, 169, 170, 181
  variola 39
  vesicular stomatitis 145
  West Nile 39, 144
  yellow fever 39

W

Warts 158, 171
  virus 175
Washing, in virus diagnosis 28–37
  labware 31
  glassware 31, 32
  ultrasonic cleaning 32
West Nile virus, cytopathic effect of 39, 144

Y

Yeast infection 138
Yellow fever 39, 159, 160
Yolk sac 44
  inoculating 52, 53
  harvesting